CITES Orchid Checklist

Volume 2

For the genera:

Cymbidium, *Dendrobium* (selected sections only), *Disa*, *Dracula* and *Encyclia*

T0141413

Compiled by:

Jacqueline A Roberts, Lee R Allman, Clive R Beale, Richard W Butter, Kevin R Crook & H Noel McGough

Assisted by a selected panel of orchid experts

Royal Botanic Gardens, Kew

© Copyright 1997
The Trustees of The Royal Botanic Gardens Kew

First published in 1997

General editor of series: Jacqueline A Roberts

ISBN 1 900347 34 2

Produced with the financial assistance of
the CITES Nomenclature Committee,
the Royal Botanic Gardens, Kew and the
American Orchid Society

Cover design by Media Resources RBG Kew
Printed and bound by The Stationery Office, Norwich

The *American Orchid Society* is very pleased to have been able to participate in the publication of a versatile and important document such as the CITES Orchid Checklist (Volume 2). This marks a milestone for the *American Orchid Society*, now in its eighth decade, as it is the first time we have contributed as a funding institution from our newly created Conservation Fund. While the *American Orchid Society* has occasionally funded select conservation-oriented projects in the past, the financial support given here should serve to underline anew the commitment of the *American Orchid Society* to international orchid conservation efforts.

L'*American Orchid Society* est très heureuse d'avoir pu participer à la publication d'un document aussi important et souple d'emploi que la Liste CITES des orchidées (volume 2). Cette action est un jalon dans l'activité de l'*American Orchid Society*, qui en est maintenant à sa huitième décennie, car pour la première fois, nous avons agi en tant qu'institution de financement et alloué une contribution par le biais de notre Fonds pour la conservation nouvellement créé. Alors que par le passé l'*American Orchid Society* a occasionnellement financé des projets axés sur la conservation, l'appui financier fourni ici souligne à nouveau notre engagement dans l'action menée au plan international en faveur de la conservation des orchidées.

La *American Orchid Society* tiene el honor de haber participado en la publicación de una obra tan útil e importante como la Lista de orquídeas de la CITES (Volumen 2). Este hecho constituye un acontecimiento sin precedentes para la *American Orchid Society*, ya que es la primera vez desde su creación, hace más de 80 años, que actúa como institución de financiamiento con cargo a nuestro recién creado Fondo para la Conservación. Si bien la *American Orchid Society* había financiado ocasionalmente de conservación concretos, el apoyo financiero prestado en esta ocasión servirá para subrayar una vez más el compromiso de la *American Orchid Society* en favor de la conservación de las orquídeas a nivel internacional.

NED NASH
Director of Education and Conservation
American Orchid Society
September 1997

FOREWORD

The number of orchid species forms half the total of all species currently included in the CITES appendices, although not all orchids are traded internationally. However, in order to have a clear understanding of the trade and of the possible impacts on the natural wild populations, it is important that standardized nomenclature is available. The Secretariat therefore welcomes the publication of the second volume on orchid names, and it hopes that the work on the nomenclature of the remaining taxa in international trade will be completed soon. This is particularly important for the development of projects under the planned Review of Significant Trade in Plants included in Appendix II, agreed by the Conference of the Parties at its 10th meeting.

In the foreword to the first volume of the Orchid Checklist, I referred to the improvements in CITES implementation for plants and the active work of the Plants Committee. In the past few years the number of Parties and non-governmental organizations participating in its meetings has steadily increased. They not only listen to the discussions but actively participate as well, for all participants have their own responsibility in the large area of activities covered by CITES.

Here, I would in particular like to mention the *American Orchid Society*. They provided a substantial contribution of USD 10,000 for the publication of the second volume of the CITES Orchid Checklist. This support should not be seen only as a welcome financial contribution but also as a reflection of an ever-improving co-operation between producers and consumers, both working for the conservation of the species in the wild.

IZGREV TOPKOV
Secretary General, CITES
September 1997

AVANT-PROPOS

Les espèces d'orchidées constituent la moitié des espèces actuellement inscrites aux annexes CITES; quoi qu'il en soit, toutes les orchidées ne font pas l'objet d'un commerce international. Pour bien comprendre le commerce et ses effets possibles sur les populations d'orchidées dans la nature, il est important qu'une nomenclature normalisée soit disponible. Le Secretariat se félicite donc de la publication du second volume sur les noms des orchidées et espère que le travail sur la nomenclature des taxons restants présents dans le commerce international sera bientôt terminé. C'est particulièrement important pour l'élaboration de projects dans le cadre de l'étude du commerce important de plantes inscrites à l'Annexe II, décidé par la Conference des Parties à sa 10ᵉ session.

Dans l'avant-propos du premier volume de la liste des orchidées, je mentionnais les améliorations dans la mise en oeuvre de la CITES concernant les plantes, et le travail actif du Comité pour les plantes. Ces dernières années, le nombre de Parties et d'organizations non gouvernementales participatant aux sessions du Comité a régulièrement augmenté. Non seulement suivent-ils les dénbats mais encore y participent-ils activement; en effet, chaque participant a ses propres responsabilités dans la vaste gamme d'activités couvertes par la CITES.

Je tiens à citer ici en particulier l'*American Orchid Society*, qui a fourni une contribution substantielle de USD 10 000 pour la publication du second volume de la liste CITES des orchidées. Cet appui ne doit pas être considéré simplement comme une contribution financière bienvenue mais comme le reflet d'une cooperation toujours meilleure entre producteurs et consommateurs en vue de la conservation des espèces dans la nature.

IZGREV TOPKOV
Secretary General, CITES
September 1997

PROLOGO

El número de especies de orquídeas constituye la mitad del total de todas las especies incluidas en los Apéndices de la CITES, a pesar de que no todas ellas son objeto de comercio internacional. No obstante, a fin de tener una noción clara del comercio y de los posible impactos sobre las poblaciones en el medio silvestre es importante disponer de nomenclatura normalizada. Así, pues, la Secretaría acoge con beneplácito la publicación del segundo volumen sobre orquídeas y alberga la esperanza de que la labor sobre la nomenclatura de los taxa restantes en el comercio internacional se completará en breve, ya que es particularmente importante para la preparación de proyectos en el marco del Examen del comercio significativo de plantas incluidas en el Apéndice II, aprobado por la décima reunión de la Conferencia de las Partes.

En el prólogo del primer volumen de la Lista de orquídeas, hice referencia a la mejora en la aplicación de la CITES para las plantas y a la activa labor del Comité de Flora. En los últimos años se ha registrado un aumento considerable en lo que concierne a la participación de las Partes y organizaciones no gubernamentales en sus reuniones. Los participantes no solo asisten a las deliberaciones sino que participan activamente en las mismas, ya que todos ellos son asimismo responsables de una amplia esfera de acividades amparadas por la CITES.

Deseo aprovechar esta oportunidad para dar las gracias a la *American Orchid Society* por la generosa contribución aportada de USD 10.000 para la publicación del segundo volumen de la Lista de orquídeas CITES. Este apoyo no debería considerarse únicamente como una contribución financiera sino como un reflejo de la colaboración cada vez más intensa entre los productores y consumidores, que aunan sus esfuerzos en favor de la conservación de las especies en la naturaleza.

IZGREV TOPKOV
Secretary General, CITES
September 1997

Acknowledgements / Remerciements / Reconocimientos

The compilers would like to thank colleagues at the Royal Botanic Gardens, Kew and the following orchid experts for their help with the preparation of the checklist for publication. All suggestions for amendments were gratefully noted and included at the discretion of the compilers. We would particularly like to thank **Mr Andre Schuiteman** and **Dr Ed de Vogel** for their comprehensive review of the final version of the text, using information from the database ASORCH at the Rijksherbarium, Leiden, The Netherlands. Finally, we are very grateful to the CITES Secretariat for translating the text into French and Spanish.

Les compilateurs tiennent à remercier leurs collègues des *Royal Botanic Gardens* de Kew et les experts en orchidées suivants, pour leur aide dans la préparation de la Liste en vue de sa publication. Il a été pris note avec gratitude de toutes les suggestions, qui ont été incluses à la discrétion des compilateurs. Nous adressons des remerciements particuliers à **Mr. Andre Schuiteman** et **Dr Ed de Vogel**, qui ont examiné de manière approfondie la version finale du texte, utilisant les données de la base de données ASORCH du Rijksherbarium, Pays-Bas. Nous remercions vivement le Secrétariat CITES d'avoir traduit le texte en espagnol et en français.

Las personas encargadas de la recopilación de este volumen desean dar las gracias a sus colegas del *Royal Botanic Gardens*, Kew, así como a los siguientes especialistas en orquídeas por su ayuda en la preparación de la presente lista. Se tomó nota con agradecimiento de todas las sugerencias de enmienda, que fueron incluidas conforme a los criterios de los responsables de este volumen. En particular deseamos dar las gracias al **Mr. Andre Schuiteman** y al **Dr. Ed de Vogel** por la revisión de la versión definitiva del texto, utilizando información procedente de la base de datos ASORCH, mantenida en el Rijksherbarium, Leiden, Países Bajos. Agradecemos a la Secretaría CITES por la traducción del texto al español y al francés.

International Panel of Orchid Experts
Groupe international d'experts des orchidées
Grupo Internacional de Expertos en Orquídeas

Dr L Averyanov	Russia	Dr H Kurzweil	South Africa
Dr A T de Brito	Brazil	Dr H P Linder	South Africa
Dr S-C Chen	China	Dr C Luer	USA
Dr P J Cribb	UK	Mr E S Manning	UK
Mrs I F la Croix	UK	Mr D M Menzies	UK
Dr E Dauncey	UK	Dr G Romero	USA
Dr R Dressler	USA	Dr C L Withner	USA
Mr A Easton	New Zealand	Dr M Wolff	Germany
Mr J Hermans	UK	Dr H Wood	USA
Mr T Inskipp	UK	Mr J J Wood	UK

CONTENTS

TABLE DES MATIERES

Préambule

INDICE

Preámbulo

CITES CHECKLIST - ORCHIDACEAE

PREAMBLE

1. Background

The 1992 Conference of the Parties to the Convention on International Trade in Endangered Species of Wild Fauna and Flora (CITES) adopted Resolution Conf. 8.19 which called for the production of a standard reference to the names of Orchidaceae.

The Vice-Chairman of the CITES Nomenclature Committee was charged with the responsibility of co-ordinating the input needed to produce such a reference.

The orchid genera identified as priorities in the *Review of Significant Trade in Species of Plants included in Appendix II of CITES* (CITES Doc. 8.31) would be treated first. The checklists (or parts thereof) as they came available would be put to the Conference of the Parties for approval.

At its third meeting (Chiang Mai, Thailand, November 1992) the Plants Committee extensively discussed a proposal by the Vice-Chairman of the Nomenclature Committee regarding the possible mechanisms to develop the Standard Reference. The Plants Committee endorsed a procedure by which compilations from the available literature, made on a central database, would be circulated to a panel of international experts for consultation and final decisions on the valid names to be used for the taxa concerned. During the development of the checklists every effort was made to recruit national experts from the range states using the contact network of the regional representatives of the Plants Committee. The response was poor and it is hoped that publication of the checklists will encourage increased participation in future volumes.

Based on the recommendations of the Plants Committee and on those of Resolution Conf. 8.19, the CITES Secretariat established a Memorandum of Understanding with the Royal Botanic Gardens, Kew for the preparation of this reference. The work started on the 1 July 1993. Recommendation 6 of the *Review of Significant Trade of Species of Plants included in Appendix II of* CITES outlined the following genera as priorities:

Aerangis, Angraecum, Ascocentrum, Bletilla, Brassavola, Calanthe, Catasetum Cattleya, Coelogyne, Comparettia, Cymbidium, Cypripedium, Dendrobium, Disa, Dracula, Encyclia, Epidendrum, Laelia, Lycaste, Masdevallia, Miltonia, Miltoniopsis, Odontoglossum, Oncidium, Paphiopedilum, Paraphalaenopsis, Phalaenopsis, Phragmipedium, Renanthera, Rhynchostylis, Rossioglossum, Sophronitis, Vanda and *Vandopsis.*

The Memorandum of Understanding stated that completed and verified checklists of the following groups should be prepared for consideration by the ninth meeting of the Conference of the Parties:

Cattleya, Cypripedium, Laelia, Paphiopedilum, Phalaenopsis, Phragmipedium, Pleione and *Sophronitis.*

This work was completed in 1995, resulting in the publication of CITES Orchid Checklist Volume 1, which also included accounts of the genera *Constantia, Paraphalaenopsis* and *Sophronitella.*

Volume 1 is the adopted CITES standard to be used as a guideline when making reference to the species names of the genera concerned.

Preamble

In February 1996 a second Memorandum of Understanding stated that completed and verified checklists should be prepared for the following groups:

Cymbidium, Dendrobium (selected sections only), *Disa, Dracula* and *Encyclia.*

Given the size of the genus *Dendrobium* it was decided to concentrate on those taxa which were likely to be seen in trade. The selection of these species was based on the views of the Panel of Orchid Experts and the taxa found in trade records.

Work on these groups began immediately. The Vice-Chairman of the Nomenclature Committee reported on progress at the sixth (Puerto de la Cruz, Tenerife, July 1995) and seventh (San José, Costa Rica, November 1996) meetings of the Plants Committee. The Tenth Meeting of the Conference of the Parties to CITES approved the continuing process and adopted this volume as the guideline when making reference to the species names of the genera concerned.

2. Computer aspects

Hardware: The database was set up on a Compaq LTE Lite 4/25C laptop computer using ALICE software.

Database system: The ALICE database system was used to handle the data collection. "ALICE handles distribution, uses, common names, descriptions, habitats, synonymy, bibliography and other classes of data for species, subspecies or varieties. It allows users to design their own reports for checklists, studies, monographs or conservation lists for example.

The ALICE System is a family of programs which operate on ALICE databases. Each has a different purpose.

ALICE: The main program for the database developer: database design, data capture and editing. Reports use simple pre-defined formats.

ATEXT: To incorporate, edit and view free-text species descriptions.

AQUERY: Flexible data retrieval with a simple interface suitable even for those with only superficial knowledge of the database.

AWRITE: For designing and printing reports.

NVIEW: For nomenclatural exploration of a database.

ALEX: To export data subsets into other program formats including: SDF, dBASE, Tabular or Fixed-length fields, DELTA and XDF.

ASLICE: To export data subsets as new independent ALICE databases.

AMIE: For routine maintenance of ALICE databases.

ALICE is currently written in XBASE and C. ALICE can run under DOS, Windows, VMS, UNIX and XENIX. ALICE can also run on Oracle, Recital and Ingres." (ALICE Software Partnership, November 1991).

3. Compilation procedures

- Primary references were identified by orchid specialists based at the Royal Botanic Gardens, Kew.
- A Panel of Orchid Experts was established in order to review each stage of the checklist.
- Information was entered into the ALICE taxonomic database and a preliminary report produced.
- Preliminary reports for each genus were distributed to the Panel of Orchid Experts for their comments on any additions or amendments needed.
- Additions and amendments returned from the Panel members were entered into the database. These were linked to a reference contained in the bibliography at the end of the report on each genus. (Not included in this checklist, but copies held for reference at RBG, Kew).
- This sequence was repeated five times for each genus to allow full consultation with the Panel.
- A preliminary checklist was compiled for review at the Tenth Meeting of the Conference of the Parties to CITES (COP10).
- Dr Ed de Vogel and Dr Andre Schuiteman were consulted as external editors to review the final draft.
- Additions and amendments subsequent to COP10 were added to the database.
- Format for publication was agreed with the CITES Secretariat and reports generated using AWRITE and prepared for camera-ready copy using Microsoft Word for Windows version 6.

4. Conservation

During the consultation process with the Panel of Orchid Experts, information was also requested on the conservation status of the species concerned. Copies of the present and proposed new IUCN categories of threat were distributed to the Panel for their use. Unfortunately, there were limited returns from the Panel and the information has not been included in this checklist.

5. How to use the checklist

It is intended that this Checklist be used as a quick reference for checking accepted names, synonymy and distribution. The reference is therefore divided into three main parts:

Part I: ORCHIDACEAE BINOMIALS IN CURRENT USAGE - ALL NAMES
An alphabetical list of all accepted names and synonyms included in this checklist - a total of 3407 names (1234 accepted and 2173 synonyms).

Part II: ORCHIDACEAE BINOMIALS IN CURRENT USAGE - ACCEPTED NAMES
Separate lists for each genus. Each list is ordered alphabetically by the accepted name and details are given on current synonyms and distribution.

Part III: COUNTRY CHECKLIST
Accepted names from all genera included in this checklist are ordered alphabetically under country of distribution.

6. Conventions employed in parts I, II and III

a) Accepted names are presented in **bold roman** type.
    Synonyms are presented in *italic* type.

b) Where a synonym occurs twice, but refers to different accepted names, eg, *Angraecum crumenatum,* for both **Dendrobium crumenatum** and

Dendrobium papilioniferum, the name with an asterix is the species most likely to be encountered in trade. For example:

All Names	Accepted Name
Angraecum album-minus	**Dendrobium papilioniferum var. ephemerum**
Angraecum crumenatum	**Dendrobium crumenatum**
Angraecum crumenatum	**Dendrobium papilioniferum***
Angraecum purpureum	**Dendrobium papilioniferum**

*Species most likely to be in trade (in this example, **Dendrobium papilioniferum**).

c)   Where an accepted name and a synonym are the same, but refer to different species, eg, **Cymbidium atropurpureum** and *Cymbidium atropurpureum* (**Cymbidium aloifolium**), the name with an asterix is the species most likely to be seen in trade. For example:

All Names	Accepted Name
Cymbidium aspidistrifolium	**Cymbidium lancifolium**
Cymbidium atropurpureum	
Cymbidium atropurpureum	**Cymbidium aloifolium***

*Species most likely to be in trade (in this example, **Cymbidium aloifolium**).

NB: In examples b) and c) it is necessary to double-check by reference to the distribution as detailed in Part II. For instance, in the example c), if it was known that the plant in question came from The Philippines it is likely to be **Cymbidium atropurpureum** as **Cymbidium aloifolium** is not found there.

d)   Natural hybrids have been included in the checklist and are indicated by the multiplication sign ×. They are arranged alphabetically within the lists.

7. Number of names entered for each genus:
Cymbidium (Accepted: 67, Synonyms: 230); *Dendrobium* (Accepted: 692, Synonyms: 1225); *Disa* (Accepted: 155, Synonyms 159); *Dracula* (Accepted: 105, Synonyms: 103); *Encyclia* (Accepted: 216, Synonyms 456).

8. Geographical areas
Country names follow the United Nations standard as laid down in Country Names. *Terminology Bulletin* August 1995. United Nations 347:1-41.

9. Orchidaceae controlled by CITES
The family Orchidaceae is listed on Appendix II of CITES. In addition the following taxa are listed on Appendix I at time of publication:

Cattleya trianaei
Dendrobium cruentum
Laelia jongheana
Laelia lobata
Paphiopedilum spp.
Peristeria elata
Phragmipedium spp.

Renanthera imschootiana
Vanda coerulea

10. Bibliography
Primary reference sources used in the compilation of checklists:

Carnevali, G. & Ramírez, I. (1988). *Revision del género Encyclia.* Bol. Comit. Orquideología Soc. Venez. Cien. Nat. 23: 13-87.

Cribb, P.J. & Wood, J.J. (1994). *A Checklist of the Orchids of Borneo.* Royal Botanic Gardens Kew, UK.

Cribb, P.J. (1986). *A Revision of the Antelope and "Latourea" Dendrobiums.* Royal Botanic Gardens, Kew, UK.

Dauncey, E.A. (1994). *Towards a Revision of Dendrobium Sw. Section Pedilonum Blume (Orchidaceae).* Unpublished PhD thesis of the University of Reading, UK.

Dressler, R.L. & Pollard, G.E. (1974). *The Genus Encyclia in Mexico.* Asociation Mexicana de Orquideología, Mexico.

Du Puy, D. & Cribb, P.J. (1988). *The Genus Cymbidium.* Royal Botanic Gardens, Kew, UK.

Lewis, B.A. & Cribb, P.J. (1989). *Orchids of Vanuatu.* Royal Botanic Gardens, Kew, UK.

Lewis, B.A. & Cribb, P.J. (1991). *Orchids of the Soloman Islands and Bougainville.* Royal Botanic Gardens, Kew, UK.

Linder, H.P. (1981). Taxonomic Studies in the Disinae (Orchidaceae) IV. A Revision of Disa Berg. Section Micranthe Lindl. *Bull. Nat. Plantentuin Belg.* 51(3/4): 255-346.

Luer, C.A. (1993). *Icones Pleurothallidinarum X. Systematics of Dracula.* Missouri Botanical Garden, USA.

Pradhan, U.C. (1979). *Indian Orchids (A guide to indentification and culture)* vol 2. U.C. Pradhan, Kalimpong.

Schelpe, E.A. & Stewart, J. (1990). *Dendrobiums, an Introduction to the species in cultivation.* Orchid Sundries, Dorset, UK.

Seidenfaden, G. & Wood, J.J. (1992). *The Orchids of Peninsular Malaysia and Singapore - A revision of R. E. Holttum: Orchids of Malaya.* Fredensborg, Denmark: Olsen & Olsen.

Seidenfaden, G. (1985). Orchid Genera in Thailand XII; Dendrobium. *Opera Botanica* 83 pp295.

Upton, W.T. (1989). *Dendrobium orchids of Australia.* Portland, Oregon: Timber Press.

LISTE CITES DES ORCHIDACEAE

PREAMBULE

1. Contexte

En 1992, la Conférence des Parties à la Convention sur le commerce international des espèces de faune et de flore sauvages menacées d'extinction (CITES) a adopté la résolution Conf. 8.19 dans laquelle elle recommande la préparation d'une liste normalisée de référence des noms d'Orchidaceae.

Le vice-président du Comité CITES de la nomenclature a été chargé de coordonner les informations reçues en vue de préparer cette liste.

Les genres d'orchidées classés comme prioritaires dans l'Examen du commerce important d'espèces végétales inscrites à l'Annexe II de la CITES (document CITES Doc. 8.31) devaient être les premiers traités. Les listes (ou parties de listes) devaient être soumises à la Conférence des Parties pour approbation à mesure qu'elles seraient disponibles.

A sa troisième session (Chiang Mai, Thaïlande, novembre 1992) le Comité pour les plantes a abondamment discuté d'une proposition du vice-président du Comité de la nomenclature concernant les mécanismes possibles d'élaboration d'une liste de référence normalisée. Le Comité pour les plantes a approuvé une procédure par laquelle les compilations faites à partir de la littérature disponible, sur une base de données centrale, seraient envoyées à un groupe d'experts internationaux pour consultation et décision finale sur les noms valides devant être utilisés pour les taxons concernés. Au cours de la préparation des listes, il a été fait appel aux experts nationaux des Etats de l'aire de répartition, en utilisant le réseau de contacts des représentants régionaux du Comité pour les plantes. Toutefois, il y a eu peu de réponses. L'on espère que la publication de la présente Liste favorisera une participation accrue aux futurs volumes.

Sur la base des recommandations du Comité pour les plantes et de la résolution Conf. 8.19, le Secrétariat CITES a établi un protocole d'accord avec les *Royal Botanic Gardens* de Kew pour la préparation de la liste de référence. Le travail a débuté le 1er juillet 1993. La recommandation 6 de l'Examen du commerce important d'espèces végétales inscrites à l'Annexe II de la CITES recommande de considérer les genres suivants comme prioritaires:

Aerangis, Angraecum, Ascocentrum, Bletilla, Brassavola, Calanthe, Catasetum Cattleya, Coelogyne, Comparettia, Cymbidium, Cypripedium, Dendrobium, Disa, Dracula, Encyclia, Epidendrum, Laelia, Lycaste, Masdevallia, Miltonia, Miltoniopsis, Odontoglossum, Oncidium, Paphiopedilum, Paraphalaenopsis, Phalaenopsis, Phragmipedium, Renanthera, Rhynchostylis, Rossioglossum, Sophronitis, Vanda et *Vandopsis.*

Le protocole d'accord prévoyait que des listes seraient établies et vérifiées pour les groupes suivants, et soumises pour examen à la neuvième session de la Conférence des Parties:

Cattleya, Cypripedium, Laelia, Paphiopedilum, Phalaenopsis, Phragmipedium, Pleione et *Sophronitis.*

Cette tâche a été achevée en 1995 et a abouti à la publication de la Liste CITES des orchidées, Volume 1, qui inclut également des données sur les genres *Constantia*, *Paraphalaenopsis* et *Sophronitella*.

Le Volume 1 est la liste de référence CITES adoptée, à utiliser lorsqu'on se réfère aux noms des espèces de ces genres.

En février 1996, un second protocole d'accord a porté sur l'établissement et la vérification de listes pour les groupes suivants:

Cymbidium, Dendrobium (certaines sections sélectionnées), *Disa, Dracula* et *Encyclia*.

Compte tenu de la taille du genre *Dendrobium*, il a été décidé que l'activité serait concentrée sur les taxons les plus susceptibles d'être présents dans le commerce. Leur sélection a été fondée sur les vues du groupe d'experts des orchidées et sur les taxons figurant dans les données commerciales.

Le travail sur ces taxons a commencé sans délai. Le vice-président du Comité de la nomenclature a présenté un rapport d'activité aux sixième (Puerto de la Cruz, Tenerife, juillet 1995) et septième (San José, Costa Rica, novembre 1996) sessions du Comité pour les plantes. La 10e session de la Conférence des Parties à la CITES a approuvé la poursuite du processus et a adopté le présent volume comme ligne directrice dans l'utilisation des noms d'espèces des genres concernés.

2. Aspects informatiques

Matériel: La base de données a été créée sur un ordinateur portable Compaq LTE Lite 4/25C en utilisant le logiciel ALICE.

Système de base de données: Le système ALICE a été utilisé pour enregistrer les données. "ALICE traite la répartition géographique, les utilisations, les noms communs, les descriptions, les habitats, les synonymes, la bibliographie et d'autres catégories de données relatives aux espèces, sous-espèces ou variétés. Il permet aux utilisateurs de créer leurs propres rapports sous forme de listes, d'études, de monographies, de listes de conservation, etc.

Le système ALICE est une famille de programmes dont chacun a sa raison d'être.

ALICE: Principal programme pour la création d'une base de données: conception de la base de données, saisie et modification des données. Les rapports utilisent des configurations simples, définies préalablement.

ATEXT: Pour incorporer, modifier et consulter les descriptions d'espèces.

AQUERY: Consultation aisée des données au moyen d'un interface simple, accessible même à ceux qui n'ont qu'une connaissance superficielle des bases de données.

AWRITE: Pour créer et imprimer les rapports.

NVIEW: Pour explorer la nomenclature d'une base de données.

ALEX: Pour exporter des sous-groupes de données vers d'autres programmes, notamment: SDF, dBASE, champs séparés ou de longueur fixe, DELTA et XDF.

ASLICE: Pour exporter des sous-groupes de données en tant que nouvelles bases de données indépendantes ALICE.

AMIE: Pour les fonctions de routine des bases de données ALICE.

ALICE est actuellement écrit en XBASE et en C. Le système ALICE peut être exploité sous DOS, Windows, VMS, UNIX et XENIX, mais également sous Oracle, Recital et Ingres." (*ALICE Software Partnership*, novembre 1991).

3. Procédure de compilation
- Les principales références ont été identifiées par les spécialistes des orchidées des *Royal Botanic Gardens* de Kew.
- Un groupe d'experts des orchidées a été établi afin d'examiner la Liste à chaque étape.
- Des informations ont été entrées dans la base de données taxonomiques ALICE et un rapport préliminaire a été préparé.
- Des rapports préliminaires sur chaque genre ont été remis au groupe d'experts pour qu'ils formulent leurs commentaires sur les additions ou amendements nécessaires.
- Les additions et amendements des membres du groupe ont été entrés dans la base de données. Ils ont été reliés à une référence incluse dans la bibliographie à la fin du rapport sur chaque genre. (Non incluse dans la présente Liste mais dont des copies sont à disposition, pour référence, au RBG, à Kew).
- Le groupe a été pleinement consulté cinq fois.
- Une liste préliminaire a été compilée et soumise pour examen à la 10ᵉ session de la Conférence des Parties à la CITES (CdP 10).
- Dr. Ed De Vogel et Mr. Andre Schuiteman ont été consultés en tant que réviseurs externes et ont examiné le projet final.
- Les additions et amendements postérieurs à la CdP 10 ont été ajoutés à la base de données.
- La présentation retenue pour la publication a été convenue avec le Secrétariat CITES et les rapports ont été créés en utilisant AWRITE et préparés pour la publication en utilisant Microsoft Word pour Windows, version 6.

4. Conservation
Au cours du processus de consultation du groupe d'experts, des informations ont été demandées sur l'état de conservation des espèces concernées. Des copies des catégories actuelles et des nouvelles catégories de menaces proposées par l'UICN ont été distribuées au groupe. Malheureusement, il y a eu peu de réponses du groupe, aussi les informations n'ont-elles pas été incluses dans la présente Liste.

5. Comment utiliser la Liste?
La présente Liste devrait être utilisée comme liste de référence pour vérifier rapidement les noms reconnus, les synonymes et la répartition géographique. Elle est divisée en trois parties principales:

Première partie: BINOMES D'ORCHIDACEAE ACTUELLEMENT EN USAGE - TOUS LES NOMS
Liste alphabétique de tous les noms reconnus et des synonymes inclus dans la Liste, soit un total de 3407 noms.

Deuxième partie: BINOMES D'ORCHIDACEAE ACTUELLEMENT EN USAGE - NOMS RECONNUS

Listes séparées pour chaque genre. Chaque liste est donnée dans l'ordre alphabétique des noms reconnus et comporte des indications sur les synonymes et la répartition géographique actuels.

Troisième partie: LISTE PAR PAYS

Les noms reconnus de tous les genres inclus dans cette liste sont donnés par ordre alphabétique sous chaque pays de l'aire de répartition.

6. Conventions utilisées dans les première, deuxième et troisième parties

a)   Les noms reconnus sont en bold roman.
     Les synonymes sont en italic.

b)   Lorsque le synonyme apparaît deux fois mais renvoie à des noms reconnus différents - par exemple, Angraecum crumenatum, pour Dendrobium crumenatum et Dendrobium papilioniferum, le nom comportant un astérisque est celui de l'espèce la plus susceptible d'être trouvée dans le commerce. Exemple:

Tous les noms	**Nom reconnu**
Angraecum album-minus	**Dendrobium papilioniferum** var. **ephemerum**
Angraecum crumenatum	**Dendrobium crumenatum**
Angraecum crumenatum	**Dendrobium papilioniferum***
Angraecum purpureum	**Dendrobium papilioniferum**

*Espèce la plus susceptible d'être trouvée dans le commerce (dans cet exemple, **Dendrobium papilioniferum**)

c)   Lorsque le nom reconnu et un synonyme sont les mêmes mais renvoient à des espèces différentes - par exemple **Cymbidium atropurpureum** et *Cymbidium atropurpureum* (**Cymbidium aloifolium**), le nom comportant un astérisque est celui de l'espèce la plus susceptible d'être trouvée dans le commerce. Exemple:

Tous les noms	**Nom reconnu**
Cymbidium aspidistrifolium	**Cymbidium lancifolium**
Cymbidium atropurpureum	
Cymbidium atropurpureum	**Cymbidium aloifolium***

*Espèce la plus susceptible d'être trouvée dans le commerce (dans cet exemple, **Cymbidium aloifolium**)

NB: Dans les exemples b) et c), il faut effectuer une double vérification en se référant à la répartition géographique indiquée dans la deuxième partie. Ainsi, dans l'exemple c), si l'on sait que la plante vient de les Philippines, il s'agit probablement de **Cymbidium atropurpureum** car **Cymbidium aloifolium** ne se trouve pas dans ce pays.

d)   Les hybrides naturels figurent dans les listes, dans l'ordre alphabétique, et sont indiqués par le signe de multiplication ×.

7.   Nombre de noms entrés pour chaque genre:
Cymbidium (reconnus: 67, synonymes: 230); *Dendrobium* (reconnus: 692, synonymes: 1225); *Disa* (reconnus: 155, synonymes 159); *Dracula* (reconnus: 105, synonymes: 103); *Encyclia* (reconnus: 216, synonymes 456).

8.   Régions géographiques
Les noms des pays sont ceux figurant dans le *Terminology Bulletin No. 347* des Nations Unies, de mai 1995.

9.   Orchidées soumises aux contrôles CITES
La famille des Orchidaceae est inscrite à l'Annexe II de la CITES. De plus, les taxons suivants étaient inscrits à l'Annexe I au moment de la publication de la Liste:

Cattleya trianaei
Dendrobium cruentum
Laelia jongheana
Laelia lobata
Paphiopedilum spp.
Peristeria elata
Phragmipedium spp.
Renanthera imschootiana
Vanda coerulea

10. Bibliographie
Principales sources de références utilisées dans la compilation des listes:

Carnevali, G. & Ramírez, I. (1988).   *Revision del género Encyclia.*   Bol. Comit. Orquideología Soc. Venez. Cien. Nat. 23: 13-87.

Cribb, P.J. & Wood, J.J. (1994).   *A Checklist of the Orchids of Borneo.*   Royal Botanic Gardens Kew, UK.

Cribb, P.J. (1986).   *A Revision of the Antelope and "Latourea" Dendrobiums.*   Royal Botanic Gardens, Kew, UK.

Dauncey, E.A. (1994).   *Towards a Revision of Dendrobium Sw. Section Pedilonum Blume (Orchidaceae).*   Unpublished PhD thesis of the University of Reading, UK.

Dressler, R.L. & Pollard, G.E. (1974).   *The Genus Encyclia in Mexico.*   Asociation Mexicana de Orquideología, Mexico.

Du Puy, D. & Cribb, P.J. (1988).   *The Genus Cymbidium.*   Royal Botanic Gardens, Kew, UK.

Lewis, B.A. & Cribb, P.J. (1989).   *Orchids of Vanuatu.*   Royal Botanic Gardens, Kew, UK.

Lewis, B.A. & Cribb, P.J. (1991). *Orchids of the Soloman Islands and Bougainville.* Royal Botanic Gardens, Kew, UK.

Linder, H.P. (1981). Taxonomic Studies in the Disinae (Orchidaceae) IV. A Revision of Disa Berg. Section Micranthe Lindl. *Bull. Nat. Plantentuin Belg.* 51(3/4): 255-346.

Luer, C.A. (1993). *Icones Pleurothallidinarum X. Systematics of Dracula.* Missouri Botanical Garden, USA.

Pradhan, U.C. (1979). *Indian Orchids (A guide to indentification and culture)* vol 2. U.C. Pradhan, Kalimpong.

Schelpe, E.A. & Stewart, J. (1990). *Dendrobiums, an Introduction to the species in cultivation.* Orchid Sundries, Dorset, UK.

Seidenfaden, G. & Wood, J.J. (1992). *The Orchids of Peninsular Malaysia and Singapore - A revision of R. E. Holttum: Orchids of Malaya.* Fredensborg, Denmark: Olsen & Olsen.

Seidenfaden, G. (1985). Orchid Genera in Thailand XII; Dendrobium, *Opera Botanica* 83 pp295.

Upton, W.T. (1989). *Dendrobium orchids of Australia.* Portland, Oregon: Timber Press.

LISTA CITES - ORCHIDACEAE

PREAMBULO

1. Antecedentes

En 1992 la Conferencia de las Partes en la Convención sobre el Comercio Internacional de Especies Amenazadas de Fauna y Flora Silvestres (CITES) aprobó la Resolución Conf. 8.19, en la que se solicita que se prepare una referencia normalizada sobre los nombres de las Orchidaceae.

Se encargó al Vicepresidente del Comité de Nomenclatura de la CITES que coordinase las tareas necesarias para preparar dicha referencia.

Se acordó abordar en primer lugar los géneros de orquídeas identificados como prioritarios en el Examen del comercio signifitativo de especies incluidas en el Apéndice II de la CITES (CITES Doc. 8.31). Las listas (o partes de las mismas) se presentarían a la aprobación de la Conferencia de las Partes a medida que se fuesen preparando.

En su tercera reunión (Chiang Mai, Tailandia, noviembre de 1992), el Comité de Flora examinó detenidamente una propuesta del Vicepresidente del Comité de Nomenclatura sobre los posibles mecanismos para preparar un referencia normalizada. El Comité de Flora ratificó un procedimiento mediante el cual se efectuarían recopilaciones a partir de la literatura disponible, que se introducirían en una base central de datos, y se presentaría a un grupo internacional de expertos para efectuar consultas y tomar decisiones sobre los nombres que deberían utilizarse para los taxa en cuestión. Durante la preparación de las listas se desplegaron esfuerzos para contactar a expertos nacionales de los Estados del área de distribución, utilizando como coordinadores a los representantes regionales ante el Comité de Flora. La respuesta no fue tan positiva como se había previsto y se espera que la publicación de las listas alentará la participación en futuros volúmenes.

En virtud de las recomendaciones del Comité de Flora y las que figuran en la Resolución Conf. 8.19, la Secretaría de la CITES estableció un Memorándum de Entendimiento con el Royal Botanic Gardens, Kew, para preparar dicha referencia. Las tareas se iniciaron el 1 de julio de 1993. En la Recomendación 6 del Examen del comercio significativo de especies de plantas incluidas en el Apéndice II de la CITES, se hace hincapié en dar prioridad a los siguientes géneros:

Aerangis, Angraecum, Ascocentrum, Bletilla, Brassavola, Calanthe, Catasetum Cattleya, Coelogyne, Comparettia, Cymbidium, Cypripedium, Disa, Dendrobium, Dracula, Encyclia, Epidendrum, Laelia, Lycaste, Masdevallia, Miltonia, Miltoniopsis, Odontoglossum, Oncidium, Paphiopedilum, Paraphalaenopsis, Phalaenopsis, Phragmipedium, Renanthera, Rhynchostylis, Rossioglossum, Sophronitis, Vanda y *Vandopsis.*

En el Memorándum de Entendimiento se indicaba que deberían prepararse listas completas y verificadas de los grupos siguientes para su consideración por la novena reunión de la Conferencia de las Partes:

Cattleya, Cypripedium, Laelia, Paphiopedilum, Phalaenopsis, Phragmipedium, Pleione y *Sophronitis.*

Esta labor se completó en 1995, dando como resultado la publicación del Volumen 1 de la Lista de orquídeas de la CITES, en el que se incluyen también los géneros *Constantia, Paraphalaenopsis* y *Sophronitella.*

El Volumen 1 es la referencia CITES adoptada para su utilización como directriz al referirse a los nombres de las especies de los generos en cuestión.

En febrero de 1996 se estableció un segundo Memorándum de Entendimiento en el que se dejaba constancia de que se prepararían listas de control completas y verificadas de los grupos siguientes:

Cymbidium, Dendrobium (únicamente las secciones seleccionadas), *Disa, Dracula* y *Encyclia.*

Habida cuenta del tamaño del género *Dendrobium*, se tomó la decisión de concentrarse en aquellos taxa que podrían ser objeto de comercio. La selección de dichas especies se efectuó en base a las opiniones expresadas por el Grupo de expertos sobre orquídeas y los taxa que figuran en los registros sobre comercio.

Las tareas sobre estos grupos se inició inmediatamente. El Vicepresidente del Comité de Nomenclatura informó sobre los progresos realizados en las reuniones sexta y séptima del Comité de Flora (Puerto de la Cruz, Tenerife, julio de 1995) y (San José, Costa Rica, noviembre de 1996). En la décima reunión de la Conferencia de las Partes se aprobó que continuara este proceso y se adoptó el presente volumen como directriz al hacer referencia a los nombres de las especies de los generos de que se trata.

2. Programa informático

Hardware: La base de datos se instaló en un Compaq LTE Lite 4/25C computadora portátil equipada con un software ALICE.

Sistema de base de datos: Se utilizó el sistema de base de datos ALICE para la recopilación de los datos. ALICE se ocupa de la distribución, utilización, nombres comunes, descripciones, hábitat, sinónimos, bibliografía y otras clases de datos para las especies, subespecies o variedades. Permite a los usuarios preparar sus propios informes para las listas, estudios, monografías o listas de conservación.

El sistema ALICE es un conjunto de programas que operan en la base de datos ALICE. Cada uno de ellos tiene una finalidad diferente.

ALICE: El programa principal para el desarrollo de la base de datos: concepción de base de datos, acopio de datos y edición. Para los informes se utilizan formatos sencillos definidos con antelación.

ATEXT: Para incorporar, editar y consultar descripciones de especies en texto libre.

AQUERY: Recuperación flexible de datos con un interfaz simple apropiado incluyo para aquellos que no disponen de gran conocimiento sobre la base de datos.

AWRITE: Para diseñar e imprimir informes.

NVIEW: Para consultar la nomenclatura en la base de datos.

ALEX: Para enviar subconjuntos de datos a otros programas: SDF, dBASE, Tabular o Fixed-length fields, DELTA y XDF.

ASLICE: Para enviar subconjuntos de datos como una nueva base de datos ALICE independiente.

AMIE: Para el mantenimiento de rutina de la base de datos ALICE.

ALICE opera actualmente en XBASE y C, pero puede funcionar en DOS, Windows, VMS, UNIX y XENIX, así como en Oracle, Recital e Ingres." (ALICE Software Partnership, noviembre de 1991).

3. Procedimiento para la recopilación

- Las referencias preliminares fueron identificadas por especialistas en orquídeas del Royal Botanic Gardens, Kew.
- Se estableció un Grupo de expertos sobre orquídeas para que revisara cada una de las fases de la lista.
- Se introdujo la información en la base de datos taxonómica de ALICE y se preparó un informe preliminar.
- Los informes preliminares respecto de cada género se distribuyeron al Grupo de expertos sobre orquídeas para que formulase comentarios sobre cualquier adición o enmienda.
- Las adiciones y enmiendas remitidas por el Grupo de expertos se introdujeron en la base de datos, vinculándolas a una referencia contenida en la bibliografía al final del informe sobre cada género. (No incluida en la presente lista de control, pero se guardan copias para referencia en el RBG, Kew).
- Esta secuencia se repitió cinco veces para cada género, a fin de realizar consultas pormenorizadas con el Grupo.
- Se compiló una lista preliminar para presentar a la consideración de la décima reunión de la Conferencia de las Partes en la CITES (COP10).
- El Dr. Ed De Vogel y el Mr. Andre Schuiteman participaron como editores externos para revisar la versión final.
- Después de la COP10 se añadieron adiciones y enmiendas a la base de datos.
- El formato para la publicación se acordó con la Secretaría de la CITES, los informes se efectuaron en AWRITE y el material preparado para la cámara se preparó utilizando Microsoft Word for Windows, versión 6.

4. Conservación

Durante el proceso de consultas con el Grupo de expertos sobre orquídeas, se solicitó también información sobre el estado de conservación de las especies en cuestión. Se transmitió al Grupo copia de las categorías de amenaza de la UICN actuales y propuestas. Lamentablemente, se recibieron pocas respuestas del Grupo de expertos y por tanto la información no se incluyó en la presente lista.

5. Cómo emplear esta lista

La idea es que esta lista se utilice como referencia rápida para controlar los nombres aceptados, los sinónimos y la distribución. Así, pues, la referencia se divide en tres partes principales:

Parte I: ORCHIDACEAE BINOMIALES UTILIZADOS NORMALMENTE - TODOS LOS NOMBRES

Una lista por orden alfabético de todos los nombres y sinónimos aceptados - un total de 3407 nombres.

Parte II: ORCHIDACEAE BINOMIALES UTILIZADOS NORMALMENTE - NOMBRES ACEPTADOS

Listas separadas para cada género. En cada lista se presentan por orden alfabético los nombres aceptados, con información sobre los sinónimos y la distribución.

Parte III: LISTA POR PAISES

Los nombres aceptados para todos los géneros incluidos en esta lista se presentan por orden alfabético según el país de distribución.

6. **Sistema de presentación utilizado en las Partes I, II y III**
   a) Los nombres aceptados se presentan en tipo de letra negrita y romano.
   Los sinónimos se presentan en letra cursiva.

   b) Cuando un sinónimo aparece dos veces, pero se refiere a diferentes nombres aceptados, a saber, *Angraecum crumenatum*, para **Dendrobium crumenatum** y **Dendrobium papilioniferum**, el nombre acompañado de un asterisco se refiere a la especie que con mayor probabilidad se encontrará en el comercio. Por ejemplo:

Todos los nombres	Nombre aceptado
Angraecum album-minus	**Dendrobium papilioniferum var. ephemerum**
Angraecum crumenatum	**Dendrobium crumenatum**
Angraecum crumenatum	**Dendrobium papilioniferum***
Angraecum purpureum	**Dendrobium papilioniferum**

   *La especie que con mayor probabilidad se encontrará en el comercio (en este ejemplo, **Dendrobium papilioniferum**).

   c) Cuando un nombre aceptado es igual al sinónimo, pero se refiere a especies diferentes, a saber, **Cymbidium atropurpureum** y *Cymbidium atropurpureum* (**Cymbidium aloifolium**), el nombre acompañado por un asterisco se refiere a la especie que con mayor probabilidad se encontrará en el comercio. Por ejemplo:

Todos los nombres	Nombre aceptado
Cymbidium aspidistrifolium	**Cymbidium lancifolium**
Cymbidium atropurpureum	
Cymbidium atropurpureum	**Cymbidium aloifolium***

   *La especie que con mayor probabilidad se encontrará en el comercio (en este ejemplo, **Cymbidium aloifolium**).

   NB: En los ejemplos b) y c) es preciso efectuar doble verificación en lo que concierne a la distribución, como se indica en la Parte II. Por ejemplo, en el caso c), si se sabe que la planta en cuestión procede de las Filipinas, es probable que se trate de **Cymbidium atropurpureum** ya que **Cymbidium aloifolium** no se encuentra en ese país.

   d) En la lista se han incluido los híbridos naturales y se indican con el signo de multiplicar "×". Se presentan por orden alfabético.

7. **Número de nombres incluidos para cada género:**
Cymbidium (Aceptados: 67, Sinónimos: 230); *Dendrobium* (Aceptados: 692, Sinónimos: 1225); *Disa* (Aceptados: 155, Sinónimos: 159); *Dracula* (Aceptados: 105, Sinónimos: 103); *Encyclia* (Aceptados: 216, Sinónimos: 456).

8. Areas geográficas

Para los nombres de los países se ha seguido la referencia oficial de las Naciones Unidas. *Terminology Bulletin*, United Nations, No.347, 1995.

9. Orchidaceae controladas por la CITES

La familia de Orchidaceae está incluida en el Apéndice II de la CITES. Además, en el momento de esta publicación, están incluidos en el Apéndice I los siguientes taxa:

Cattleya trianaei
Dendrobium cruentum
Laelia jongheana
Laelia lobata
Paphiopedilum spp.
Peristeria elata
Phragmipedium spp.
Renanthera imschootiana
Vanda coerulea

10. Bibliografía

Principales fuentes de referencia utilizadas para la recopilación de las listas:

Carnevali, G. & Ramírez, I. (1988). *Revision del género Encyclia.* Bol. Comit. Orquideología Soc. Venez. Cien. Nat. 23: 13-87.

Cribb, P.J. & Wood, J.J. (1994). *A Checklist of the Orchids of Borneo.* Royal Botanic Gardens Kew, UK.

Cribb, P.J. (1986). *A Revision of the Antelope and "Latourea" Dendrobiums.* Royal Botanic Gardens, Kew, UK.

Dauncey, E.A. (1994). *Towards a Revision of Dendrobium Sw. Section Pedilonum Blume (Orchidaceae).* Unpublished PhD thesis of the University of Reading, UK.

Dressler, R.L. & Pollard, G.E. (1974). *The Genus Encyclia in Mexico.* Asociation Mexicana de Orquideología, Mexico.

Du Puy, D. & Cribb, P.J. (1988). *The Genus Cymbidium.* Royal Botanic Gardens, Kew, UK.

Lewis, B.A. & Cribb, P.J. (1989). *Orchids of Vanuatu.* Royal Botanic Gardens, Kew, UK.

Lewis, B.A. & Cribb, P.J. (1991). *Orchids of the Soloman Islands and Bougainville.* Royal Botanic Gardens, Kew, UK.

Linder, H.P. (1981). Taxonomic Studies in the Disinae (Orchidaceae) IV. A Revision of Disa Berg. Section Micranthe Lindl. *Bull. Nat. Plantentuin Belg.* 51(3/4): 255-346.

Luer, C.A. (1993). *Icones Pleurothallidinarum X. Systematics of Dracula.* Missouri Botanical Garden, USA.

Pradhan, U.C. (1979). *Indian Orchids (A guide to indentification and culture)* vol 2. U.C. Pradhan, Kalimpong.

Schelpe, E.A. & Stewart, J. (1990). *Dendrobiums, an Introduction to the species in cultivation.* Orchid Sundries, Dorset, UK.

Seidenfaden, G. & Wood, J.J. (1992). *The Orchids of Peninsular Malaysia and Singapore - A revision of R. E. Holttum: Orchids of Malaya.* Fredensborg, Denmark: Olsen & Olsen.

Seidenfaden, G. (1985). Orchid Genera in Thailand XII; Dendrobium. *Opera Botanica* 83 pp295.

Upton, W.T. (1989). *Dendrobium orchids of Australia.* Portland, Oregon: Timber Press.

Part I: ORCHIDACEAE BINOMIALS IN CURRENT USAGE
Ordered alphabetically on All Names for the genera:

Cymbidium, Dendrobium (selected sections only*), Disa, Dracula* and *Encyclia*

Première partie: BINOMES D'ORCHIDACEAE ACTUELLEMENT EN USAGE
Par ordre alphabétique de tous les noms pour les genre:

Cymbidium, Dendrobium (certaines sections sélectionnées), *Disa, Dracula* et *Encyclia*

Parte I: ORCHIDACEAE BINOMIALES UTILIZADOS NORMALMENTE
Presentados por orden alfabético: todos los nombres para el genero:

Cymbidium, Dendrobium (únicamente las secciones seleccionadas), *Disa, Dracula* y *Encyclia*

ALPHABETICAL LISTING OF ALL NAMES FOR THE GENERA:
Cymbidium, Dendrobium (selected sections only), *Disa, Dracula* and *Encyclia*

LISTES ALPHABETIQUES DE TOUS LES NOMS POUR LES GENRE:
Cymbidium, Dendrobium (certaines sections sélectionnées), *Disa, Dracula* et *Encyclia*

PRESENTACION POR ORDEN ALFABETICO DE TODOS LOS NOMBRES
PARA EL GENERO:
Cymbidium, Dendrobium (únicamente las secciones seleccionadas), *Disa, Dracula* y *Encyclia*

ALL NAMES TOUS LES NOMS TODOS LOS NOMBRES	ACCEPTED NAME NOM RECONNU NOMBRES ACEPTADOS
Aerides borassii	**Cymbidium aloifolium**
Amblostoma tridactylum var. *mexicanum*	**Encyclia glauca**
Amphigena leptostachys	**Disa tenuis**
Amphigena tenuis	**Disa tenuis**
Anacheilium alagoense	**Encyclia alagonsis**
Anacheilium allemanii	**Encyclia allemanii**
Anacheilium allemanoides	**Encyclia allemanoides**
Anacheilium caetense	**Encyclia caetensis**
Anacheilium calamarium	**Encyclia calamaria**
Anacheilium campos-portoi	**Encyclia campos-portoi**
Anacheilium cochleatum	**Encyclia cochleata**
Anacheilium cochleatum subsp. *arrogans*	**Encyclia cochleata**
Anacheilium faresianum	**Encyclia faresiana**
Anacheilium faustum	**Encyclia fausta**
Anacheilium fragrans	**Encyclia fragrans**
Anacheilium glumaceum	**Encyclia glumacea**
Anacheilium gramatoglossum	**Encyclia grammatoglossa**
Anacheilium hartwegii	**Encyclia hartwegii**
Anacheilium inversum	**Encyclia inversa**
Anacheilium kautskyi	**Encyclia kautskyi**
Anacheilium lividum	**Encyclia livida**
Anacheilium moojenii	**Encyclia moojenii**
Anacheilium papilio	**Encyclia papilio**
Anacheilium radiatum	**Encyclia radiata**
Anacheilium suzanense	**Encyclia suzanensis**
Anacheilium vespa	**Encyclia vespa**
Anacheilium widgrenii	**Encyclia widgrenii**
Angraecum album-minus	**Dendrobium papilioniferum** var. **ephemerum**
Angraecum crumenatum	**Dendrobium crumenatum***
Angraecum crumenatum	**Dendrobium papilioniferum**
Angraecum purpureum	**Dendrobium capituliflorum**
Angraecum purpureum var. *sylvestre*	**Dendrobium purpureum**
Aphyllorchis aberrans	**Cymbidium macrorhizon**
Aporum aciculare	**Dendrobium aciculare**
Aporum acinaciforme	**Dendrobium acinaciforme**
Aporum aloifolium	**Dendrobium aloifolium**
Aporum anceps	**Dendrobium anceps**
Aporum anceps	**Dendrobium nathanielis**
Aporum babiense	**Dendrobium babiense**
Aporum banaense	**Dendrobium acinaciforme**
Aporum bicornutum	**Dendrobium bicornutum**

*For explanation see page 4, point 6
*Voir les explications page 9, point 6
*Para mayor explicación, véase la página 15, point 6

ALL NAMES	ACCEPTED NAMES
Aporum blumei	Dendrobium blumei
Aporum calceolum	Dendrobium calceolum
Aporum chalandei	Dendrobium bowmanii
Aporum cinnabarinum	Dendrobium cinnabarinum
Aporum clavator	Dendrobium clavator
Aporum cochinchinense	Dendrobium aloifolium
Aporum compressimentum	Dendrobium compressimentum
Aporum crucilabre	Dendrobium crucilabre
Aporum crumenatum	Dendrobium crumenatum
Aporum cuspidatum	Dendrobium nathanielis
Aporum cymbulipes	Dendrobium cymbulipes
Aporum dalatense	Dendrobium dalatense
Aporum eboracense	Dendrobium eboracense
Aporum ephemerum	Dendrobium papilioniferum var. ephemerum
Aporum equitans	Dendrobium ventricosum
Aporum goldfinchii	Dendrobium goldfinchii
Aporum gracile	Dendrobium gracile
Aporum grande	Dendrobium grande
Aporum grootingsii	Dendrobium grootingsii
Aporum hainanense	Dendrobium hainanense
Aporum hendersonii	Dendrobium hendersonii
Aporum heterocaulon	Dendrobium exile
Aporum hymenocentrum	Dendrobium hymenocentrum
Aporum incrassatum	Dendrobium indivisum
Aporum incurvociliatum	Dendrobium incurvociliatum
Aporum indivisum	Dendrobium indivisum
Aporum jenkinsii	Dendrobium parciflorum
Aporum junceum	Dendrobium junceum
Aporum koeteianum	Dendrobium aciculare
Aporum korthalsii	Dendrobium korthalsii
Aporum kwashotense	Dendrobium crumenatum
Aporum lawiense	Dendrobium lawiense
Aporum leonis	Dendrobium leonis
Aporum litorale	Dendrobium litorale
Aporum lobatum	Dendrobium lobatum
Aporum lobulatum	Dendrobium lobulatum
Aporum mannii	Dendrobium mannii
Aporum micranthum	Dendrobium aloifolium
Aporum nycteridoglossum	Dendrobium nycteridoglossum
Aporum obcordatum	Dendrobium obcordatum
Aporum papilio	Dendrobium papilio
Aporum papilioniferum	Dendrobium papilioniferum
Aporum peculiare	Dendrobium setifolium
Aporum planibulbe	Dendrobium planibulbe
Aporum platyphyllum	Dendrobium nycteridoglossum
Aporum prostratum	Dendrobium prostratum
Aporum pseudocalceolum	Dendrobium pseudocalceolum
Aporum reflexitepalum	Dendrobium reflexitepalum
Aporum rhodostele	Dendrobium rhodostele
Aporum rivesii	Dendrobium chryseum
Aporum rosellum	Dendrobium rosellum
Aporum roxburghii	Dendrobium calceolum
Aporum sagittatum	Dendrobium sagittatum
Aporum serra	Dendrobium acinaciforme
Aporum serra	Dendrobium aloifolium

*For explanation see page 4, point 6
*Voir les explications page 9, point 6
*Para mayor explicación, véase la página 15, point 6

21

ALL NAMES	ACCEPTED NAMES
Aporum setifolium	**Dendrobium setifolium**
Aporum smithianum	**Dendrobium smithianum**
Aporum strigosum	**Dendrobium tetraedre**
Aporum subulatoides	**Dendrobium subulatoides**
Aporum tenellum	**Dendrobium tenellum**
Aporum tenue	**Dendrobium tenue**
Aporum tetraedre	**Dendrobium tetraedre**
Aporum tetralobum	**Dendrobium tetralobum**
Aporum tricuspis	**Dendrobium tricuspe**
Aporum truncatum	**Dendrobium truncatum**
Aporum uncatum	**Dendrobium uncatum**
Aporum verlaquii	**Dendrobium terminale**
Arethusantha bletioides	**Cymbidium elegans**
Aulizeum cochleatum	**Encyclia cochleata**
Aulizeum pygmaeum	**Encyclia pygmaea**
Aulizeum variegatum	**Encyclia vespa**
Australorchis monophylla	**Dendrobium monophyllum**
Australorchis schneiderae	**Dendrobium schneiderae**
Bletia ensiformis	**Encyclia cyperifolia**
Bletia ensiformis	**Encyclia ensiformis**
Bletia nipponica	**Cymbidium macrorhizon**
Bulbophyllum lichenastrum	**Dendrobium lichenastrum**
Bulbophyllum occidentale	**Encyclia polybulbon**
Bulbophyllum oncidiochilum	**Dendrobium bifalce**
Bulbophyllum prenticei	**Dendrobium prenticei**
Bulbophyllum toressae	**Dendrobium toressae**
Callista acicularis	**Dendrobium aciculare**
Callista acinaciformis	**Dendrobium acinaciforme**
Callista acrobatica	**Dendrobium capillipes**
Callista adae	**Dendrobium adae**
Callista aduncum	**Dendrobium aduncum**
Callista aemula	**Dendrobium aemulum**
Callista aggregata	**Dendrobium lindleyi**
Callista albosanguinea	**Dendrobium albosanguineum**
Callista aloefolia	**Dendrobium aloifolium**
Callista alpestris	**Dendrobium monticola**
Callista amabilis	**Dendrobium amabile**
Callista amabilis	**Dendrobium hercoglossum***
Callista amethystoglossa	**Dendrobium amethystoglossum**
Callista amoena	**Dendrobium amoenum**
Callista anceps	**Dendrobium anceps**
Callista angulata	**Dendrobium podagraria**
Callista annamensis	**Dendrobium hercoglossum**
Callista anosma	**Dendrobium anosmum**
Callista antennata	**Dendrobium antennatum**
Callista aphylla	**Dendrobium aphyllum**
Callista aquea	**Dendrobium aqueum**
Callista arachnites	**Dendrobium dickasonii**
Callista aurantiaca	**Dendrobium chryseum**
Callista aurea	**Dendrobium heterocarpum**
Callista aurorosea	**Dendrobium nudum**
Callista bairdiana	**Dendrobium fellowsii**
Callista barbatula	**Dendrobium barbatulum**
Callista beckleri	**Dendrobium schoeninum**
Callista bensoniae	**Dendrobium bensoniae**
Callista bicamerata	**Dendrobium bicameratum**

*For explanation see page 4, point 6
*Voir les explications page 9, point 6
*Para mayor explicación, véase la página 15, point 6

ALL NAMES	ACCEPTED NAMES
Callista bicaudata	**Dendrobium bicaudatum**
Callista bifalcis	**Dendrobium bifalce**
Callista biflora	**Dendrobium gemellum**
Callista bigibba	**Dendrobium bigibbum**
Callista bolboflora	**Dendrobium bicameratum**
Callista boothii	**Dendrobium blumei**
Callista borneoensis	**Dendrobium aloifolium**
Callista boxallii	**Dendrobium gratiosissimum**
Callista brachypus	**Dendrobium brachypus**
Callista breviflora	**Dendrobium bicameratum**
Callista brymeriana	**Dendrobium brymerianum**
Callista bursigera	**Dendrobium secundum**
Callista calceolum	**Dendrobium calceolum**
Callista calophylla	**Dendrobium calophyllum**
Callista canaliculata	**Dendrobium canaliculatum**
Callista capillipes	**Dendrobium capillipes**
Callista carinifera	**Dendrobium cariniferum**
Callista chrysantha	**Dendrobium chrysanthum**
Callista chrysocrepis	**Dendrobium chrysocrepis**
Callista chrysotoxa	**Dendrobium chrysotoxum**
Callista chrysotoxum var. *delacourii*	**Dendrobium chrysotoxum**
Callista ciliata	**Dendrobium venustum**
Callista clavata	**Dendrobium chryseum**
Callista clavipes	**Dendrobium truncatum**
Callista cornuta	**Dendrobium hasseltii**
Callista crassinodis	**Dendrobium pendulum**
Callista crepidata	**Dendrobium crepidatum**
Callista cretacea	**Dendrobium cretaceum**
Callista crocata	**Dendrobium crocatum**
Callista cruenta	**Dendrobium cruentum**
Callista crumenata	**Dendrobium crumenatum**
Callista cucumerinum	**Dendrobium cucumerinum**
Callista cumulata	**Dendrobium cumulatum**
Callista cunninghamii	**Dendrobium cunninghami**
Callista cuspidata	**Dendrobium cuspidatum**
Callista dearei	**Dendrobium dearei**
Callista densiflora	**Dendrobium densiflorum**
Callista denudans	**Dendrobium denudans**
Callista devoniana	**Dendrobium devonianum**
Callista dicupha	**Dendrobium affine**
Callista dixantha	**Dendrobium dixanthum**
Callista draconis	**Dendrobium draconis**
Callista eriaeflora	**Dendrobium eriiflorum**
Callista erosa	**Dendrobium erosum**
Callista eulophota	**Dendrobium indivisum**
Callista fairfaxii	**Dendrobium fairfaxii**
Callista falconeri	**Dendrobium falconeri**
Callista falcorostris	**Dendrobium falcorostrum**
Callista farmeri	**Dendrobium farmeri**
Callista fimbriata	**Dendrobium blumei**
Callista findlayana	**Dendrobium findlayanum**
Callista fitzgeraldii	**Dendrobium × superbiens**
Callista flavidula	**Dendrobium spegidoglossum**
Callista foelschei	**Dendrobium foelschei**
Callista formosa	**Dendrobium formosum**
Callista fytchiana	**Dendrobium fytchianum**

*For explanation see page 4, point 6
*Voir les explications page 9, point 6
*Para mayor explicación, véase la página 15, point 6

23

ALL NAMES	ACCEPTED NAMES
Callista gibsonii	**Dendrobium gibsonii**
Callista glossotis	**Dendrobium catillare**
Callista gordonii	**Dendrobium macrophyllum**
Callista gouldii	**Dendrobium gouldii**
Callista gracilicaulis	**Dendrobium gracilicaule**
Callista gracilis	**Dendrobium gracile**
Callista graminifolia	**Dendrobium graminifolium**
Callista grandis	**Dendrobium grande**
Callista gratiosissima	**Dendrobium gratiosissimum**
Callista griffthiana	**Dendrobium griffithianum**
Callista harveyana	**Dendrobium harveyanum**
Callista hasseltii	**Dendrobium hasseltii**
Callista herbacea	**Dendrobium herbaceum**
Callista hercoglossa	**Dendrobium hercoglossum**
Callista heterocarpa	**Dendrobium heterocarpum**
Callista heyneana	**Dendrobium heyneanum**
Callista hookeriana	**Dendrobium hookerianum**
Callista hornei	**Dendrobium hornei**
Callista hymenophylla	**Dendrobium hymenophyllum**
Callista hymenoptera	**Dendrobium hymenopterum**
Callista incrassata	**Dendrobium indivisum**
Callista incurva	**Dendrobium incurvum**
Callista indivisa	**Dendrobium indivisum**
Callista infundibulum	**Dendrobium infundibulum**
Callista japonica	**Dendrobium moniliforme**
Callista jenkinsii	**Dendrobium parciflorum**
Callista juncea	**Dendrobium junceum**
Callista kingiana	**Dendrobium kingianum**
Callista kuhlii	**Dendrobium hasseltii**
Callista lamellata	**Dendrobium lamellatum**
Callista lasioglossa	**Dendrobium lasioglossum**
Callista lawana	**Dendrobium crepidatum**
Callista leonis	**Dendrobium leonis**
Callista leucochlora	**Dendrobium leucochlorum**
Callista leucolophota	**Dendrobium affine**
Callista lilacina	**Dendrobium swartzii**
Callista linawiana	**Dendrobium linawianum**
Callista linguiformis	**Dendrobium linguaeforme**
Callista lituiflora	**Dendrobium lituiflorum**
Callista lobata	**Dendrobium lobatum**
Callista loddigesii	**Dendrobium loddigesii**
Callista longicornis	**Dendrobium longicornu**
Callista lowii	**Dendrobium lowii**
Callista lubbersiana	**Dendrobium williamsonii**
Callista lucens	**Dendrobium lucens**
Callista lunata	**Dendrobium lunatum**
Callista macfarlanei	**Dendrobium johnsoniae**
Callista macrantha	**Dendrobium macranthum**
Callista macrophylla	**Dendrobium anosmum**
Callista macropus	**Dendrobium macropus**
Callista macrostachya	**Dendrobium macrostachyum**
Callista marginata	**Dendrobium xanthophlebium**
Callista marmorata	**Dendrobium marmoratum**
Callista micrantha	**Dendrobium aloifolium**
Callista microbolbon	**Dendrobium microbolbon**
Callista mirbeliana	**Dendrobium mirbelianum**

*For explanation see page 4, point 6
*Voir les explications page 9, point 6
*Para mayor explicación, véase la página 15, point 6

24

ALL NAMES	ACCEPTED NAMES
Callista moniliformis	Dendrobium moniliforme
Callista monophylla	Dendrobium monophyllum
Callista moorei	Dendrobium moorei
Callista mortii	Dendrobium mortii
Callista moschata	Dendrobium moschatum
Callista moulmeinensis	Dendrobium moulmeinense
Callista mutabilis	Dendrobium mutabile
Callista nathanielis	Dendrobium nathanielis
Callista nobilis	Dendrobium nobile
Callista nuda	Dendrobium nudum
Callista ochreata	Dendrobium ochreatum
Callista oculata	Dendrobium fimbriatum
Callista ophioglossa	Dendrobium smillieae
Callista ovata	Dendrobium ovatum
Callista pachyglossa	Dendrobium pachyglossum
Callista palpebrae	Dendrobium palpebrae
Callista pandurifera	Dendrobium panduriferum
Callista parca	Dendrobium parcum
Callista parishii	Dendrobium parishii
Callista pendula	Dendrobium pendulum
Callista phalaenopsis	Dendrobium phalaenopsis
Callista porphyrochila	Dendrobium porphyrochilum
Callista primulina	Dendrobium primulinum
Callista pugioniformis	Dendrobium pugioniforme
Callista pulchella	Dendrobium pulchellum
Callista purpurea	Dendrobium purpureum
Callista pychnostachya	Dendrobium pychnostachyum
Callista pygmaea	Dendrobium peguanum
Callista radians	Dendrobium radians
Callista ramosa	Dendrobium ramosum
Callista reinwardtii	Dendrobium purpureum
Callista rhodopterygia	Dendrobium rhodopterygium
Callista rigida	Dendrobium rigidum
Callista sanguinolenta	Dendrobium sanguinolentum
Callista sarcantha	Dendrobium cuspidatum
Callista scabrillinguis	Dendrobium scabrilingue
Callista scortechinii	Dendrobium anosmum
Callista sculpta	Dendrobium sculptum
Callista secunda	Dendrobium secundum
Callista senilis	Dendrobium senile
Callista smillieae	Dendrobium smillieae
Callista spatella	Dendrobium acinaciforme
Callista speciosa	Dendrobium speciosum
Callista spectabilis	Dendrobium spectabile
Callista stratiotes	Dendrobium stratiotes
Callista strebloceras	Dendrobium strebloceras
Callista striolata	Dendrobium striolatum
Callista strongylantha	Dendrobium strongylanthum
Callista stuartii	Dendrobium stuartii
Callista stuposa	Dendrobium stuposum
Callista suavissima	Dendrobium chrysotoxum
Callista subacaulis	Dendrobium subacaule
Callista sulcata	Dendrobium sulcatum
Callista superbiens	Dendrobium × superbiens
Callista tattonianum	Dendrobium tattonianum
Callista taurina	Dendrobium taurinum

*For explanation see page 4, point 6
*Voir les explications page 9, point 6
*Para mayor explicación, véase la página 15, point 6

ALL NAMES	ACCEPTED NAMES
Callista tenella	**Dendrobium tenellum**
Callista teretifolia	**Dendrobium teretifolium**
Callista terminalis	**Dendrobium terminale**
Callista tetraedris	**Dendrobium tetraedre**
Callista tetragona	**Dendrobium tetragonum**
Callista tetrodon	**Dendrobium tetrodon**
Callista thyrsodes	**Dendrobium thyrsodes**
Callista tortilis	**Dendrobium tortile**
Callista transparens	**Dendrobium transparens**
Callista tricuspis	**Dendrobium tricuspis**
Callista trigonopus	**Dendrobium trigonopus**
Callista truncata	**Dendrobium truncatum**
Callista tuberifera	**Dendrobium planibulbe**
Callista uncata	**Dendrobium uncatum**
Callista undulata	**Dendrobium discolor**
Callista veitchiana	**Dendrobium macrophyllum**
Callista veratrifolia	**Dendrobium lineale**
Callista vexans	**Dendrobium hercoglossum**
Callista virginea	**Dendrobium virgineum**
Callista wardiana	**Dendrobium wardianum**
Callista wattii	**Dendrobium wattii**
Callista williamsonii	**Dendrobium williamsonii**
Callista xanthophlebia	**Dendrobium xanthophlebium**
Cattleya citrina	**Encyclia citrina**
Cattleya karwinskii	**Encyclia citrina**
Ceraia simplicissima	**Dendrobium crumenatum**
Coelandria smillieae	**Dendrobium smillieae**
Coelogyne triptera	**Encyclia pygmaea**
Cymbidium aberrans	**Cymbidium macrorhizon**
Cymbidium acuminatum	**Cymbidium ensifolium** subsp. **haematodes**
Cymbidium acutum	**Cymbidium dayanum**
Cymbidium affine	**Cymbidium mastersii**
Cymbidium albo-jucundissimum	**Cymbidium sinense**
Cymbidium albo-marginatum	**Cymbidium ensifolium**
Cymbidium alborubens	**Cymbidium dayanum**
Cymbidium albuciflorum	**Cymbidium madidum**
Cymbidium aliciae	**Cymbidium cyperifolium**
Cymbidium aloifolium*	
Cymbidium aloifolium	**Cymbidium bicolor**
Cymbidium aloifolium	**Cymbidium bicolor** var. **pubescens**
Cymbidium aloifolium	**Cymbidium finlaysonianum**
Cymbidium aloifolium var. *pubescens*	**Cymbidium bicolor** var. **pubescens**
Cymbidium angustifolium	**Cymbidium dayanum**
Cymbidium aphyllum	**Cymbidium macrorhizon**
Cymbidium aphyllum	**Dendrobium aphyllum***
Cymbidium arrogans	**Cymbidium ensifolium**
Cymbidium aspidistrifolium	**Cymbidium lancifolium**
Cymbidium atropurpureum	
Cymbidium atropurpureum	**Cymbidium aloifolium***
Cymbidium atropurpureum var. *olivaceum*	**Cymbidium atropurpureum**
Cymbidium babae	**Cymbidium cochleare**
Cymbidium bambusifolium	**Cymbidium lancifolium**
Cymbidium banaense	
Cymbidium bicolor*	
Cymbidium bicolor	**Cymbidium bicolor** subsp. **obtusum**

*For explanation see page 4, point 6
*Voir les explications page 9, point 6
*Para mayor explicación, véase la página 15, point 6

ALL NAMES	ACCEPTED NAMES
Cymbidium bicolor subsp. **bicolor**	
Cymbidium bicolor subsp. **obtusum**	
Cymbidium bicolor subsp. **pubescens**	
Cymbidium borneense	
Cymbidium canaliculatum	
Cymbidium canaliculatum var. *barrettii*	Cymbidium canaliculatum
Cymbidium canaliculatum var. *marginatum*	Cymbidium canaliculatum
Cymbidium canaliculatum var. *sparkesii*	Cymbidium canaliculatum
Cymbidium carnosum	Cymbidium cyperifolium
Cymbidium caulescens	Cymbidium lancifolium
Cymbidium celebicum	Cymbidium bicolor subsp. pubescens
Cymbidium cerinum	Cymbidium faberi
Cymbidium chinense	Cymbidium sinense
Cymbidium chloranthum	
Cymbidium chuen-lan	Cymbidium goeringii
Cymbidium cochleare	
Cymbidium cordigerum	Encyclia cordigera
Cymbidium crassifolium	Cymbidium bicolor subsp. obtusum
Cymbidium cuspidatum	Cymbidium lancifolium
Cymbidium cyperifolium	
Cymbidium cyperifolium	Cymbidium faberi var. szechuanicum
Cymbidium cyperifolium subsp. **arrogans**	
Cymbidium cyperifolium subsp. **cyperifolium**	
Cymbidium dayanum	
Cymbidium dayanum var. *austro-japonicum*	Cymbidium dayanum
Cymbidium defoliatum	
Cymbidium densiflorum	Cymbidium elegans
Cymbidium devonianum	
Cymbidium diurnum	Encyclia diurna
Cymbidium eburneum	
Cymbidium eburneum var. *austro-japonicum*	Cymbidium dayanum
Cymbidium eburneum var. *dayana*	Cymbidium dayanum
Cymbidium eburneum var. *dayi*	Cymbidium eburneum
Cymbidium eburneum var. *obtusum*	Cymbidium eburneum
Cymbidium eburneum var. *parishii*	Cymbidium parishii
Cymbidium eburneum var. *philbrickianum*	Cymbidium eburneum
Cymbidium ecristatum	Cymbidium ensifolium
Cymbidium elegans	
Cymbidium elegans var. *lutescens*	Cymbidium elegans
Cymbidium elegans var. *obcordatum*	Cymbidium elegans
Cymbidium elongatum	
Cymbidium ensifolium	
Cymbidium ensifolium	Cymbidium ensifolium subsp. haematodes
Cymbidium ensifolium	Cymbidium sinense
Cymbidium ensifolium subsp. **ensifolium**	
Cymbidium ensifolium subsp. **haematodes**	
Cymbidium ensifolium var. *arrogans*	Cymbidium ensifolium
Cymbidium ensifolium var. *estriatum*	Cymbidium ensifolium
Cymbidium ensifolium var. *haematodes*	Cymbidium ensifolium subsp. haematodes
Cymbidium ensifolium var. *misericors*	Cymbidium ensifolium
Cymbidium ensifolium var. *munronianum*	Cymbidium munronianum
Cymbidium ensifolium var. *munronianum*	Cymbidium sinense
Cymbidium ensifolium var. *striatum*	Cymbidium ensifolium
Cymbidium ensifolium var. *yakibaran*	Cymbidium ensifolium

*For explanation see page 4, point 6
*Voir les explications page 9, point 6
*Para mayor explicación, véase la página 15, point 6

Part I: All Names / Tous les Noms / Todos los Nombres

ALL NAMES	ACCEPTED NAMES
Cymbidium erectum	Cymbidium aloifolium
Cymbidium erythraeum	
Cymbidium erythrostylum	
Cymbidium erythrostylum var. *magnificum*	Cymbidium erythrostylum
Cymbidium faberi	
Cymbidium faberi var. faberi	
Cymbidium faberi var. *omeiense*	Cymbidium kanran
Cymbidium faberi var. szechuanicum	
Cymbidium finlaysonianum	
Cymbidium finlaysonianum var. *atropurpureum*	Cymbidium atropurpureum
Cymbidium flaccidum	Cymbidium bicolor subsp. obtusum
Cymbidium floribundum	
Cymbidium floribundum var. *pumilum*	Cymbidium floribundum
Cymbidium formosanum	Cymbidium goeringii
Cymbidium formosanum var. *gracillimum*	Cymbidium goeringii var. serratum
Cymbidium forrestii	Cymbidium goeringii
Cymbidium fragrans	Cymbidium sinense
Cymbidium fukienense	Cymbidium faberi
Cymbidium gibsonii	Cymbidium lancifolium
Cymbidium giganteum var. *hookerianum*	Cymbidium hookerianum
Cymbidium giganteum var. *lowianum*	Cymbidium lowianum
Cymbidium giganteum var. *wilsonii*	Cymbidium wilsonii
Cymbidium giganticum	Cymbidium iridioides
Cymbidium goeringii	
Cymbidium goeringii var. *arrogans*	Cymbidium goeringii var. serratum
Cymbidium goeringii var. goeringii	
Cymbidium goeringii var. longibracteatum	
Cymbidium goeringii var. serratum	
Cymbidium goeringii var. tortisepalum	
Cymbidium gomphocarpum	Cymbidium suave
Cymbidium gongshanense	
Cymbidium gonzalesii	Cymbidium ensifolium
Cymbidium gracillimum	Cymbidium goeringii var. serratum
Cymbidium grandiflorum	Cymbidium hookerianum
Cymbidium grandiflorum var. *kalawensis*	Cymbidium lowianum var. iansonii
Cymbidium grandiflorum var. *punctatum*	Cymbidium hookerianum
Cymbidium gyokuchin	Cymbidium ensifolium
Cymbidium gyokuchin var. *soshin*	Cymbidium ensifolium
Cymbidium gyokuchin var. *arrogans*	Cymbidium ensifolium
Cymbidium hartinahianum	
Cymbidium hennisianum	Cymbidium erythraeum
Cymbidium hilii	Cymbidium canaliculatum
Cymbidium hookerianum	
Cymbidium hookerianum var. *lowianum*	Cymbidium lowianum
Cymbidium hoosai	Cymbidium sinense
Cymbidium illiberale	Cymbidium floribundum
Cymbidium insigne	
Cymbidium insigne var. *album*	Cymbidium insigne
Cymbidium insigne var. *sanderi*	Cymbidium insigne
Cymbidium intermedium	Cymbidium aloifolium
Cymbidium iridifolium	Cymbidium madidum
Cymbidium iridioides	
Cymbidium javanicum	Cymbidium lancifolium
Cymbidium javanicum var. *aspidistrifolium*	Cymbidium lancifolium
Cymbidium javanicum var. *pantlingii*	Cymbidium lancifolium
Cymbidium kanran	

*For explanation see page 4, point 6
*Voir les explications page 9, point 6
*Para mayor explicación, véase la página 15, point 6

ALL NAMES

Cymbidium kanran var. *babae*
Cymbidium kanran var. *latifolium*
Cymbidium kanran var. *misericors*
Cymbidium kerrii
Cymbidium kinabaluense
Cymbidium koran
Cymbidium lancifolium
Cymbidium lancifolium var. *aspidistrifolium*
Cymbidium lancifolium var. *syunitianum*
Cymbidium leachianum
Cymbidium leai
Cymbidium leroyi
Cymbidium linearisepalum
Cymbidium longibracteatum

Cymbidium longifolium
Cymbidium lowianum
Cymbidium lowianum var. *concolor*
Cymbidium lowianum var. *flaveolum*
Cymbidium lowianum var. **iansonii**
Cymbidium lowianum var. **lowianum**
Cymbidium lowianum var. *superbissimum*
Cymbidium lowianum var. *viride*
Cymbidium mackinnoni
Cymbidium maclehoseae
Cymbidium macrorhizon
Cymbidium madidum
Cymbidium maguanense
Cymbidium mannii
Cymbidium mastersii
Cymbidium mastersii var. *album*
Cymbidium micans
Cymbidium micromeson
Cymbidium misericors
Cymbidium misericors var. *oreophyllum*
Cymbidium moschatum
Cymbidium munronianum
Cymbidium nagifolium
Cymbidium nanulum
Cymbidium nipponicum
Cymbidium niveo-marginatum
Cymbidium oiwakensis
Cymbidium omeiense
Cymbidium oreophyllum
Cymbidium ovatum
Cymbidium papuanum
Cymbidium parishii
Cymbidium pedicellatum
Cymbidium pendulum
Cymbidium pendulum
Cymbidium pendulum
Cymbidium pendulum
Cymbidium pendulum var. *atropurpureum*
Cymbidium pendulum var. *brevilabre*
Cymbidium pendulum var. *purpureum*
Cymbidium poilanei

ACCEPTED NAMES

Cymbidium cochleare
Cymbidium kanran
Cymbidium ensifolium
Cymbidium lancifolium

Cymbidium ensifolium

Cymbidium lancifolium
Cymbidium lancifolium
Cymbidium dayanum
Cymbidium madidum
Cymbidium madidum
Cymbidium kanran
Cymbidium goeringii var. longibracteatum
Cymbidium elegans

Cymbidium lowianum
Cymbidium lowianum

Cymbidium lowianum
Cymbidium lowianum
Cymbidium goeringii
Cymbidium lancifolium

Cymbidium bicolor subsp. obtusum

Cymbidium mastersii
Cymbidium ensifolium
Cymbidium mastersii
Cymbidium ensifolium
Cymbidium kanran
Dendrobium moschatum

Cymbidium lancifolium

Cymbidium macrorhizon
Cymbidium ensifolium
Cymbidium faberi
Cymbidium kanran
Cymbidium kanran
Dendrobium ovatum
Cymbidium lancifolium

Cymbidium macrorhizon
Cymbidium aloifolium*
Cymbidium bicolor subsp. obtusum
Cymbidium dayanum*
Cymbidium finlaysonianum
Cymbidium atropurpureum
Cymbidium finlaysonianum
Cymbidium atropurpureum
Cymbidium dayanum

*For explanation see page 4, point 6
*Voir les explications page 9, point 6
*Para mayor explicación, véase la página 15, point 6   29

ALL NAMES	ACCEPTED NAMES
Cymbidium pseudovirens	**Cymbidium goeringii**
Cymbidium pubescens	**Cymbidium bicolor** subsp. **pubescens**
Cymbidium pubescens var. *celebicum*	**Cymbidium bicolor** subsp. **pubescens**
Cymbidium pulchellum	**Cymbidium chloranthum**
Cymbidium pulchellum	**Cymbidium madidum***
Cymbidium pulchellum var. *leroyi*	**Cymbidium madidum**
Cymbidium pulchellum var. *sanderae*	**Cymbidium sanderae**
Cymbidium pulcherrimum	**Cymbidium dayanum**
Cymbidium pumilum	**Cymbidium floribundum**
Cymbidium purpureo-hiemale	**Cymbidium kanran**
Cymbidium qiubeiense	
Cymbidium queeneanum	**Cymbidium madidum**
Cymbidium rectum	
Cymbidium robustum	**Cymbidium lancifolium**
Cymbidium roseum	
Cymbidium rubrigemmum	**Cymbidium ensifolium**
Cymbidium sagamiense	**Cymbidium macrorhizon**
Cymbidium sanderae	
Cymbidium sanderi	**Cymbidium insigne**
Cymbidium sanguineolentum	**Cymbidium chloranthum**
Cymbidium sanguineum	**Cymbidium chloranthum**
Cymbidium scabroserrulatum	**Cymbidium faberi**
Cymbidium schroederi	
Cymbidium serratum	**Cymbidium goeringii** var. **serratum**
Cymbidium shimaran	**Cymbidium ensifolium**
Cymbidium siamense	**Cymbidium ensifolium** subsp. **haematodes**
Cymbidium sigmoideum	
Cymbidium sikkimense	**Cymbidium devonianum**
Cymbidium simonsianum	**Cymbidium dayanum**
Cymbidium simulans	**Cymbidium aloifolium**
Cymbidium sinense	
Cymbidium sinense var. *albo-jucundissimum*	**Cymbidium sinense**
Cymbidium sinense var. *album*	**Cymbidium sinense**
Cymbidium sinense var. *arrogans*	**Cymbidium sinense**
Cymbidium sinense var. *bellum*	**Cymbidium sinense**
Cymbidium sinokanran	**Cymbidium kanran**
Cymbidium sinokanran var. *atropurpureum*	**Cymbidium kanran**
Cymbidium sparkesii	**Cymbidium canaliculatum**
Cymbidium suave	
Cymbidium suavissimum	
Cymbidium sundaicum	**Cymbidium ensifolium** subsp. **haematodes**
Cymbidium sundaicum var. *estriata*	**Cymbidium ensifolium** subsp. **haematodes**
Cymbidium sutepense	**Cymbidium dayanum**
Cymbidium syringodorum	**Cymbidium eburneum**
Cymbidium syunitianum	**Cymbidium lancifolium**
Cymbidium szechuanensis	**Cymbidium macrorhizon**
Cymbidium szechuanicum	**Cymbidium faberi** var. **szechuanicum**
Cymbidium tentyozanense	**Cymbidium goeringii***
Cymbidium tentyozanense	**Cymbidium kanran**
Cymbidium tigrinum	
Cymbidium tortisepalum	**Cymbidium goeringii** var. **tortisepalum**
Cymbidium tortisepalum var. *viridiflorum*	**Cymbidium goeringii** var. **tortisepalum**
Cymbidium tosyaense	**Cymbidium kanran**

*For explanation see page 4, point 6
*Voir les explications page 9, point 6
*Para mayor explicación, véase la página 15, point 6

30

ALL NAMES	ACCEPTED NAMES
Cymbidium tracyanum	
Cymbidium tricolor	Cymbidium finlaysonianum
Cymbidium tsukengensis	Cymbidium goeringii var. tortisepalum
Cymbidium uniflorum	Cymbidium goeringii
Cymbidium variciferum	Cymbidium chloranthum
Cymbidium virens	Cymbidium goeringii
Cymbidium virescens	Cymbidium goeringii
Cymbidium viridiflorum	Cymbidium cyperifolium
Cymbidium wallichii	Cymbidium finlaysonianum
Cymbidium wenshanense	
Cymbidium whiteae	
Cymbidium wilsonii	
Cymbidium xiphiifolium	Cymbidium ensifolium
Cymbidium yakibaran	Cymbidium ensifolium
Cymbidium yunnanense	Cymbidium goeringii
Cyperorchis babae	Cymbidium cochleare
Cyperorchis cochlearis	Cymbidium cochleare
Cyperorchis eburnea	Cymbidium eburneum
Cyperorchis elegans	Cymbidium elegans
Cyperorchis elegans var. *blumei*	Cymbidium elegans
Cyperorchis erythrostyla	Cymbidium erythrostylum
Cyperorchis grandiflora	Cymbidium hookerianum
Cyperorchis hennisiana	Cymbidium erythraeum
Cyperorchis insignis	Cymbidium insigne
Cyperorchis longifolia	Cymbidium erythraeum
Cyperorchis lowiana	Cymbidium lowianum
Cyperorchis mastersii	Cymbidium mastersii
Cyperorchis parishii	Cymbidium parishii
Cyperorchis rosea	Cymbidium roseum
Cyperorchis schroederi	Cymbidium schroederi
Cyperorchis sigmoidea	Cymbidium sigmoideum
Cyperorchis tigrina	Cymbidium tigrinum
Cyperorchis traceyana	Cymbidium tracyanum
Cyperorchis wallichii	Cymbidium cyperifolium
Cyperorchis whiteae	Cymbidium whiteae
Cyperorchis wilsonii	Cymbidium wilsonii
Dendrobium aberrans	
Dendrobium achillis	Dendrobium calcaratum
Dendrobium aciculare	
Dendrobium acinaciforme	
Dendrobium aclinia	Dendrobium incurvum
Dendrobium actinomorphum	Dendrobium crepidatum
Dendrobium acutimentum	
Dendrobium acutisepalum	
Dendrobium adae	
Dendrobium adolphi	Dendrobium puniceum
Dendrobium aduncum	
Dendrobium aegle	Dendrobium erosum
Dendrobium aemulans	Dendrobium erosum
Dendrobium aemulum	
Dendrobium affine	
Dendrobium agathodaemonis	Dendrobium cuthbertsonii
Dendrobium aggregatum	Dendrobium lindleyi
Dendrobium aggregatum var. *jenkinsii*	Dendrobium jenkinsii
Dendrobium alabense	
Dendrobium alaticaulinum	

*For explanation see page 4, point 6
*Voir les explications page 9, point 6
*Para mayor explicación, véase la página 15, point 6

31

ALL NAMES	ACCEPTED NAMES
Dendrobium albayense	
Dendrobium albayense	**Dendrobium terminale**
Dendrobium albiviride	**Dendrobium vexillarius** var. **albiviride**
Dendrobium albiviride var. *minor*	**Dendrobium vexillarius** var. **albiviride**
Dendrobium albosanguineum	
Dendrobium alboviride	**Dendrobium linawianum**
Dendrobium album	**Dendrobium aqueum**
Dendrobium alderwereltianum	
Dendrobium alexandrae	
Dendrobium aloifolium	
Dendrobium alpestre	**Dendrobium gregulus**
Dendrobium alpestre	**Dendrobium monticola**
Dendrobium alpinum	**Dendrobium rigidifolium**
Dendrobium alterum	**Dendrobium dantaniense**
Dendrobium altomontanum	**Dendrobium caliculimentum**
Dendrobium amabile	
Dendrobium amblyogenium	
Dendrobium amethystoglossum	
Dendrobium amoenum	
Dendrobium amphigenyum	
Dendrobium anceps	
Dendrobium anceps	**Dendrobium keithii**
Dendrobium anceps	**Dendrobium leonis***
Dendrobium anceps	**Dendrobium nathanielis**
Dendrobium anceps	**Dendrobium terminale**
Dendrobium ancorarium	**Dendrobium adae**
Dendrobium andersonii	**Dendrobium draconis**
Dendrobium andreemillarae	
Dendrobium angiense	
Dendrobium angulatum	**Dendrobium podagraria**
Dendrobium angustiflorum	
Dendrobium annae	
Dendrobium annamense	
Dendrobium anosmum	
Dendrobium antelope	**Dendrobium bicaudatum**
Dendrobium antennatum	
Dendrobium anthrene	
Dendrobium apertum	
Dendrobium aphanochilum	
Dendrobium aphyllum	
Dendrobium aqueum	
Dendrobium arachnanthe	**Dendrobium discolor**
Dendrobium arachnites	**Dendrobium dickasonii**
Dendrobium arachnites	**Dendrobium unicum***
Dendrobium arachnoglossum	
Dendrobium arachnostachyum	**Dendrobium macranthum**
Dendrobium arcuatum	
Dendrobium aries	
Dendrobium aristiferum	
Dendrobium armeniacum	
Dendrobium aruanum	**Dendrobium mirbelianum**
Dendrobium ashworthiae	**Dendrobium forbesii**
Dendrobium asperifolium	**Dendrobium cuthbertsonii**
Dendrobium asphale	
Dendrobium asumburu	**Dendrobium subclausum**
Dendrobium atavus	

*For explanation see page 4, point 6
*Voir les explications page 9, point 6
*Para mayor explicación, véase la página 15, point 6

32

Part I: All Names / Tous les Noms / Todos los Nombres

ALL NAMES   **ACCEPTED NAMES**

Dendrobium atjehense
Dendrobium atractodes   Dendrobium heterocarpum
Dendrobium atromarginatum   Dendrobium cuthbertsonii
Dendrobium atroviolaceum
Dendrobium augustae-victoriae   Dendrobium lineale
Dendrobium aurantiaco-purpureum   Dendrobium prenticei
Dendrobium aurantiacum   Dendrobium chryseum
Dendrobium aurantiflavum   Dendrobium subclausum
Dendrobium aurantiroseum
Dendrobium aurantivinosum   Dendrobium brevicaule subsp.
                                         calcarium
Dendrobium aureum   Dendrobium heterocarpum
Dendrobium auriculatum
Dendrobium auroroseum   Dendrobium nudum
Dendrobium babiense
Dendrobium baeuerlenii
Dendrobium bairdianum   Dendrobium fellowsii
Dendrobium banaense   Dendrobium acinaciforme
Dendrobium barbatulum   Dendrobium ovatum
Dendrobium barbatulum*
Dendrobium baseyanum   Dendrobium calamiforme
Dendrobium basilanense
Dendrobium beckleri   Dendrobium schoeninum
Dendrobium beckleri var. *racemosum* Dendrobium racemosum
Dendrobium begoniicarpum   Dendrobium subacaule
Dendrobium begoniicarpum var. *parviflorum* Dendrobium subacaule
Dendrobium bellatulum
Dendrobium bellum   Dendrobium eximium
Dendrobium bensoniae
Dendrobium bicameratum
Dendrobium bicaudatum
Dendrobium bicornutum
Dendrobium bifalce
Dendrobium bigibbum*
Dendrobium bigibbum   Dendrobium affine
Dendrobium bigibbum var. *albomarginatum* Dendrobium × superbiens
Dendrobium bigibbum var. *georgei*   Dendrobium × lavarackianum
Dendrobium bigibbum var. *phalaenopsis* Dendrobium phalaenopsis
Dendrobium bigibbum var. *superbiens* Dendrobium × superbiens
Dendrobium bigibbum var. *superbum* Dendrobium phalaenopsis
Dendrobium bigibbum var. *venosum*   Dendrobium × lavarackianum
Dendrobium bilamellatum   Dendrobium vexillarius var. uncinatum
Dendrobium bilobulatum
Dendrobium biloculare
Dendrobium blumei*
Dendrobium blumei   Dendrobium hendersonii
Dendrobium blumei   Dendrobium planibulbe
Dendrobium bolboflorum   Dendrobium bicameratum
Dendrobium boothii   Dendrobium blumei
Dendrobium bostrychodes
Dendrobium boumaniae
Dendrobium bowmanii
Dendrobium boxallii   Dendrobium gratiosissimum
Dendrobium brachycalyptra
Dendrobium brachycentrum
Dendrobium brachyphyta   Dendrobium vexillarius var. uncinatum

*For explanation see page 4, point 6
*Voir les explications page 9, point 6
*Para mayor explicación, véase la página 15, point 6   33

ALL NAMES	ACCEPTED NAMES
Dendrobium brachypus	
Dendrobium brachythecum	**Dendrobium macrophyllum**
Dendrobium bracteosum	
Dendrobium bracteosum var. *album*	**Dendrobium bracteosum**
Dendrobium bracteosum var. *roseum*	**Dendrobium bracteosum**
Dendrobium braianense	**Dendrobium capillipes**
Dendrobium brandtiae	**Dendrobium × superbiens**
Dendrobium brassii	
Dendrobium brevicaule	
Dendrobium brevicaule subsp. **brevicaule**	
Dendrobium brevicaule subsp. **calcarium**	
Dendrobium brevicaule subsp. **pentagonum**	
Dendrobium breviflorum	**Dendrobium bicameratum**
Dendrobium brevilabium	
Dendrobium brevimentum	
Dendrobium brevimentum	**Dendrobium militare**
Dendrobium breviracemosum	**Dendrobium bifalce**
Dendrobium brinchangense	**Dendrobium hasseltii**
Dendrobium bronckartii	**Dendrobium amabile**
Dendrobium broomfieldii	**Dendrobium discolor** var. **broomfieldii**
Dendrobium brymerianum	
Dendrobium brymerianum var. **brymerianum**	
Dendrobium brymerianum var. **histrionicum**	
Dendrobium buffumii	
Dendrobium bullenianum*	
Dendrobium bullenianum	**Dendrobium taveuniense**
Dendrobium bullerianum	**Dendrobium gratiosissimum**
Dendrobium buluense	**Dendrobium mirbelianum**
Dendrobium buluense var. *kauloense*	**Dendrobium mirbelianum**
Dendrobium burbidgei	**Dendrobium bicaudatum**
Dendrobium bursigerum	**Dendrobium secundum**
Dendrobium busuangense	**Dendrobium conanthum**
Dendrobium cacatua	
Dendrobium caenosicallainum	**Dendrobium vexillarius** var. retroflexum
Dendrobium caespitificum	**Dendrobium masarangense** var. theionanthum
Dendrobium caespitosum	**Dendrobium porphyrochilum**
Dendrobium calamiforme	
Dendrobium calamiforme	**Dendrobium vagans**
Dendrobium calcaratum*	
Dendrobium calcaratum	**Dendrobium pseudorarum**
Dendrobium calcaratum subsp. **calcaratum**	
Dendrobium calcaratum subsp. **papillatum**	
Dendrobium calcariferum	
Dendrobium calcarium	**Dendrobium brevicaule** subsp. calcarium
Dendrobium calceolaria	**Dendrobium moschatum**
Dendrobium calceolum	
Dendrobium calicopis	
Dendrobium caliculimentum	
Dendrobium calophyllum	
Dendrobium calyptratum	
Dendrobium cambridgeanum	**Dendrobium ochreatum**
Dendrobium canaliculatum	
Dendrobium canaliculatum var. **canaliculatum**	

*For explanation see page 4, point 6
*Voir les explications page 9, point 6
*Para mayor explicación, véase la página 15, point 6

34

ALL NAMES	ACCEPTED NAMES
Dendrobium canaliculatum var. *nigrescens*	Dendrobium canaliculatum
Dendrobium canaliculatum var. **pallidum**	
Dendrobium canaliculatum var. *tattonianum*	Dendrobium tattonianum
Dendrobium caninum	Dendrobium crumenatum
Dendrobium capillipes	
Dendrobium capitisyork	
Dendrobium capituliflorum	
Dendrobium capituliflorum var. *viride*	Dendrobium capituliflorum
Dendrobium capra	
Dendrobium cariniferum	
Dendrobium cariniferum var. *wattii*	Dendrobium wattii
Dendrobium carrii	
Dendrobium carronii	
Dendrobium carstensziense	Dendrobium cuthbertsonii
Dendrobium caryicola	
Dendrobium castum	Dendrobium moniliforme
Dendrobium casuarinae	
Dendrobium catenatum	Dendrobium moniliforme
Dendrobium catillare	
Dendrobium catillare	Dendrobium taveuniense
Dendrobium cedricola	Dendrobium dekockii
Dendrobium cellulosum	Dendrobium sulphureum var. cellulosum
Dendrobium ceraia	Dendrobium crumenatum
Dendrobium cerasinium	Dendrobium puniceum
Dendrobium ceraula	
Dendrobium cerinum	Dendrobium sanguinolentum
Dendrobium chalandei	Dendrobium bowmanii
Dendrobium chameleon	
Dendrobium changjiangense	
Dendrobium chlorinum	Dendrobium masarangense subsp. chlorinum
Dendrobium chloroleucum	Dendrobium otaguroanum
Dendrobium chlorops	Dendrobium ovatum
Dendrobium chloropterum	Dendrobium bifalce
Dendrobium chloropterum var. *striatum*	Dendrobium bifalce
Dendrobium chlorostylum	
Dendrobium chordiforme	
Dendrobium christyanum	
Dendrobium chrysanthum	
Dendrobium chryseum	
Dendrobium chrysocephalum	Dendrobium bullenianum
Dendrobium chrysocrepis	
Dendrobium chrysoglossum	Dendrobium obtusum
Dendrobium chrysoglossum	Dendrobium pseudoglomeratum*
Dendrobium chrysolabium	Dendrobium bracteosum
Dendrobium chrysornis	Dendrobium dekockii
Dendrobium chrysotoxum	
Dendrobium chrysotoxum var. *suavissimum*	Dendrobium chrysotoxum
Dendrobium ciliatilabellum	
Dendrobium ciliatum	Dendrobium delacourii*
Dendrobium ciliatum	Dendrobium venustum
Dendrobium ciliatum var. *breve*	Dendrobium delacourii
Dendrobium ciliatum var. *rupicola*	Dendrobium venustum
Dendrobium ciliferum	Dendrobium venustum
Dendrobium cinereum	

*For explanation see page 4, point 6
*Voir les explications page 9, point 6
*Para mayor explicación, véase la página 15, point 6

Part I: All Names / Tous les Noms / Todos los Nombres

ALL NAMES	ACCEPTED NAMES
Dendrobium cinnabarinum	
Dendrobium cinnabarinum var. **angustitepalum**	
Dendrobium cinnabarinum var. **cinnabarinum**	
Dendrobium cinnabarinum var. **lamelliferum**	
Dendrobium clavator	
Dendrobium clavatum	Dendrobium chryseum
Dendrobium clavatum	Dendrobium densiflorum*
Dendrobium clavipes	Dendrobium truncatum
Dendrobium coccinellum	Dendrobium cuthbertsonii
Dendrobium cochinchinense	Dendrobium aloifolium
Dendrobium cochleatum	
Dendrobium cochliodes	
Dendrobium codonosepalum	
Dendrobium coerulescens	Dendrobium nobile*
Dendrobium coerulescens	Dendrobium putnamii
Dendrobium cogniauxianum	Dendrobium lineale
Dendrobium compactum	
Dendrobium compactum	Dendrobium gregulus
Dendrobium compressimentum	
Dendrobium compressum	Dendrobium lamellatum
Dendrobium comptonii	
Dendrobium comptonii	Dendrobium macropus
Dendrobium conanthum	
Dendrobium concavissimum	Dendrobium obtusum
Dendrobium confinale	
Dendrobium confinale	Dendrobium porphyrochilum
Dendrobium confusum	Dendrobium capituliflorum
Dendrobium conicum	
Dendrobium constrictum	
Dendrobium convexipes	
Dendrobium convolutum	
Dendrobium corallorhizon	
Dendrobium cornutum	Dendrobium hasseltii
Dendrobium crabro	
Dendrobium crassinode	Dendrobium pendulum
Dendrobium crenatifolium	
Dendrobium crepidatum	
Dendrobium crepidatum	Dendrobium loddigesii
Dendrobium crepidatum var. *avista*	Dendrobium crepidatum
Dendrobium cretaceum	
Dendrobium crispatum	Dendrobium vagans
Dendrobium crispilinguum	
Dendrobium crocatum	
Dendrobium croceocentrum	
Dendrobium crucilabre	
Dendrobium cruentum	
Dendrobium crumenatum	
Dendrobium crumenatum var. *papilioniferum*	Dendrobium papilioniferum
Dendrobium crumenatum var. *parviflorum*	Dendrobium crumenatum
Dendrobium cruttwellii	
Callista crystallina	Dendrobium crystallinum
Dendrobium crystallinum	
Dendrobium ctenoglossum	Dendrobium strongylanthum
Dendrobium cuculliferum	
Dendrobium cucullatum	Dendrobium aphyllum
Dendrobium cucumerinum	

*For explanation see page 4, point 6
*Voir les explications page 9, point 6
*Para mayor explicación, véase la página 15, point 6

ALL NAMES	ACCEPTED NAMES
Dendrobium cumulatum	
Dendrobium cunninghamii	
Dendrobium cupreum	Dendrobium moschatum
Dendrobium curtisii	Dendrobium hasseltii
Dendrobium curvicaule	
Dendrobium curviflorum	
Dendrobium curvimentum	
Dendrobium cuspidatum	
Dendrobium cuspidatum	Dendrobium nathanielis
Dendrobium cuspidatum	Dendrobium pseudocalceolum
Dendrobium cuthbertsonii	
Dendrobium cyananthum	Dendrobium hellwigianum
Dendrobium cyanocentrum	
Dendrobium cyatheicola	Dendrobium brevicaule
Dendrobium cylindricum	
Dendrobium cymboglossum	
Dendrobium cymbulipes	
Dendrobium cyperifolium	Dendrobium violaceum subsp. cyperifolium
Dendrobium d'albertisii	Dendrobium antennatum
Dendrobium dalatense	
Dendrobium dalhousieanum	Dendrobium pulchellum
Dendrobium dammerboeri	Dendrobium strebloceras
Dendrobium dantaniense	
Dendrobium daoense	
Dendrobium darjeelingensis	
Dendrobium dartoisianum	Dendrobium tortile
Dendrobium dearei	
Dendrobium dekockii	
Dendrobium delacourii*	
Dendrobium delacourii	Dendrobium venustum
Dendrobium delicatulum	
Dendrobium delicatulum	Dendrobium subacaule
Dendrobium delicatulum subsp. delicatulum	
Dendrobium delicatulum subsp. huliorum	
Dendrobium delicatulum subsp. parvulum	
Dendrobium × delicatum	
Dendrobium delphinioides	
Dendrobium deltatum	
Dendrobium demmenii	Dendrobium bicaudatum
Dendrobium dendrocolloides	
Dendrobium denneanum	Dendrobium chryseum
Dendrobium densiflorum	
Dendrobium densiflorum	Dendrobium palpebrae
Dendrobium densiflorum	Dendrobium thyrsiflorum
Dendrobium densiflorum var. *alboluteum*	Dendrobium thyrsiflorum
Dendrobium densiflorum var. *alboluteum*	Dendrobium farmeri
Dendrobium dentatum	
Dendrobium denudans	
Dendrobium denudans	Dendrobium monticola*
Dendrobium derryi	Dendrobium calicopis
Dendrobium desmotrichoides	Dendrobium rigidum
Dendrobium devonianum	
Dendrobium devosianum	
Dendrobium diceras	
Dendrobium dichaeoides*	

*For explanation see page 4, point 6
*Voir les explications page 9, point 6
*Para mayor explicación, véase la página 15, point 6

Part I: All Names / Tous les Noms / Todos los Nombres

ALL NAMES	ACCEPTED NAMES
Dendrobium dichaeoides	**Dendrobium praetermissum**
Dendrobium dichroma	
Dendrobium dickasonii	
Dendrobium dicuphum	**Dendrobium affine**
Dendrobium dicuphum var. *album*	**Dendrobium affine**
Dendrobium dicuphum var. *grandiflorum*	**Dendrobium affine**
Dendrobium dillonianum	
Dendrobium discolor	
Dendrobium discolor var. **broomfieldii**	
Dendrobium discolor var. **discolor**	
Dendrobium discolor var. **fimbrilabium**	
Dendrobium discolor var. **fuscum**	
Dendrobium discrepans	**Dendrobium puniceum**
Dendrobium dixanthum	
Dendrobium dixonianum	
Dendrobium dixsonii	**Dendrobium bracteosum**
Dendrobium dixsonii var. *eborinum*	**Dendrobium bracteosum**
Dendrobium dolichophyllum	
Dendrobium draconis	
Dendrobium drake-castilloi	**Dendrobium comptonii**
Dendrobium dryadum	**Dendrobium violaceum**
Dendrobium eboracense	
Dendrobium eburneum	**Dendrobium draconis**
Dendrobium eitapense	**Dendrobium bracteosum**
Dendrobium elobatum	**Dendrobium discolor**
Dendrobium endertii	
Dendrobium engae	
Dendrobium eoum	**Dendrobium cumulatum**
Dendrobium ephemerum	**Dendrobium papilioniferum** var. **ephemerum**
Dendrobium equitans	**Dendrobium ventricosum**
Dendrobium eriiflorum	**Dendrobium dixonianum**
Dendrobium eriiflorum*	
Dendrobium erostelle	
Dendrobium erosum	
Dendrobium erythocarpum	**Dendrobium dekockii**
Dendrobium erythroglossum	**Dendrobium falconeri**
Dendrobium erythropogon	
Dendrobium erythroxanthum	**Dendrobium bullenianum***
Dendrobium erythroxanthum	**Dendrobium taveuniense**
Dendrobium escritorii	
Dendrobium eserre	
Dendrobium eulophotum	**Dendrobium indivisum**
Dendrobium eumelinum	
Dendrobium euphues	**Dendrobium cuthbertsonii**
Dendrobium euryanthum	
Dendrobium eustachyum	**Dendrobium forbesii**
Dendrobium evaginatum	**Dendrobium aphyllum**
Dendrobium evrardii	**Dendrobium wattii**
Dendrobium exile	
Dendrobium eximium	
Dendrobium exsculptum	**Dendrobium spegidoglossum**
Dendrobium fairfaxii	
Dendrobium fairfaxii	**Dendrobium mooreanum**
Dendrobium falconeri	
Dendrobium falcorostrum	

*For explanation see page 4, point 6
*Voir les explications page 9, point 6
*Para mayor explicación, véase la página 15, point 6

38

ALL NAMES	ACCEPTED NAMES
Dendrobium fallax	Dendrobium pachyglossum
Dendrobium fantasticum	Dendrobium amphigenyum
Dendrobium fantasticum	Dendrobium woodsii
Dendrobium farmeri*	
Dendrobium farmeri	Dendrobium palpebrae
Dendrobium farmeri var. *album*	Dendrobium palpebrae
Dendrobium farmeri var. *aureoflava*	Dendrobium griffithianum
Dendrobium fellowsii	
Dendrobium ferox	Dendrobium macrophyllum
Dendrobium fesselianum	
Dendrobium filicaule	
Dendrobium fimbriatum	
Dendrobium fimbriatum var. *oculatum*	Dendrobium fimbriatum
Dendrobium fimbriatum var. *gibsonii*	Dendrobium gibsonii
Dendrobium fimbrilabium	
Dendrobium findlayanum	
Dendrobium finisterrae	
Dendrobium finisterrae var. *polystichum*	Dendrobium finisterrae
Dendrobium fitzgeraldii	Dendrobium × superbiens
Dendrobium flagellum	
Dendrobium flammula	
Dendrobium flavidulum	Dendrobium spegidoglossum
Dendrobium flaviflorum	Dendrobium chryseum
Dendrobium flavispiculum	Dendrobium cyanocentrum
Dendrobium fleckeri	
Dendrobium flexicaule	
Dendrobium floribundum	Dendrobium macropus
Dendrobium × foederatum	
Dendrobium foelschei	
Dendrobium forbesii	
Dendrobium forbesii var. *praestans*	Dendrobium forbesii
Dendrobium formosanum	Dendrobium nobile*
Dendrobium formosum	
Dendrobium fornicatum	Dendrobium obtusum
Dendrobium fractum	
Dendrobium friedericksianum	
Dendrobium friedericksianum	Dendrobium nobile*
Dendrobium friedericksianum var. *oculatum*	Dendrobium friedericksianum
Dendrobium frigidum	Dendrobium masarangense var. theionanthum
Dendrobium fruticicola	
Dendrobium fuerstenbergianum	
Dendrobium fugax	Dendrobium hendersonii
Dendrobium fulgidum	
Dendrobium fulgidum	Dendrobium cuthbertsonii*
Dendrobium fulgidum var. *angustilabre*	Dendrobium fulgidum
Dendrobium fulgidum var. **fulgidum**	
Dendrobium fulgidum var. **maritimum**	
Dendrobium fulgidum var. *purpureum*	Dendrobium cuthbertsonii
Dendrobium fulminicaule	
Dendrobium furcatopedicellatum	
Dendrobium fuscatum	Dendrobium gibsonii
Dendrobium fusiforme	Dendrobium jonesii
Dendrobium fytchianum	
Dendrobium galactanthum	Dendrobium scabrilingue
Dendrobium gamblei	Dendrobium macrostachyum

*For explanation see page 4, point 6
*Voir les explications page 9, point 6
*Para mayor explicación, véase la página 15, point 6

ALL NAMES	ACCEPTED NAMES
Dendrobium garrettii	
Dendrobium gaudens	**Dendrobium dekockii**
Dendrobium gedeanum	**Dendrobium gracile**
Dendrobium geluanum	**Dendrobium hellwigianum**
Dendrobium gemellum	
Dendrobium geminiflorum	**Dendrobium violaceum**
Dendrobium gemma	**Dendrobium masarangense** var. theionanthum
Dendrobium geotropum	
Dendrobium gibsonii	
Dendrobium giddinsii	**Dendrobium fellowsii**
Dendrobium giluwense	**Dendrobium rigidifolium**
Dendrobium glaucoviride	
Dendrobium glomeratum	
Dendrobium glomeratum	**Dendrobium pseudoglomeratum**
Dendrobium glomeriflorum	**Dendrobium catillare**
Dendrobium glossotis	**Dendrobium catillare**
Dendrobium gnomus	
Dendrobium goldfinchii	
Dendrobium goldiei	**Dendrobium × superbiens**
Dendrobium goldschmidtianum	
Dendrobium gonzalesii	**Dendrobium ceraula**
Dendrobium gordonii	**Dendrobium macrophyllum**
Dendrobium gouldii	
Dendrobium gouldii var. *acutum*	**Dendrobium gouldii**
Dendrobium gracile	
Dendrobium gracile	**Dendrobium clavator**
Dendrobium gracile	**Dendrobium linearifolium**
Dendrobium gracile	**Dendrobium setifolium**
Dendrobium gracilicaule*	
Dendrobium gracilicaule	**Dendrobium macropus**
Dendrobium gracilicaule var. *howeanum*	**Dendrobium macropus**
Dendrobium × gracillimum	
Dendrobium × gracilosum	**Dendrobium × gracillimum**
Dendrobium graminifolium	
Dendrobium grande	
Dendrobium grande	**Dendrobium keithii**
Dendrobium grantii	**Dendrobium lineale**
Dendrobium grastidioides	
Dendrobium gratiosissimum	
Dendrobium gregulus	
Dendrobium griffithianum	
Dendrobium griffithianum var. *guibertii*	**Dendrobium densiflorum**
Dendrobium × grimesii	
Dendrobium groeneveldtii	**Dendrobium cinereum**
Dendrobium grootingsii	
Dendrobium guangxiense	
Dendrobium guibertii	**Dendrobium densiflorum**
Dendrobium guilianettii	**Dendrobium mirbelianum**
Dendrobium guttatum	**Dendrobium rigidifolium**
Dendrobium gynoglottis	
Dendrobium habbemense	
Dendrobium hainanense	
Dendrobium hainanense	**Dendrobium miyakei**
Dendrobium hallieri	
Dendrobium hamaticalcar	

*For explanation see page 4, point 6
*Voir les explications page 9, point 6
*Para mayor explicación, véase la página 15, point 6

40

ALL NAMES	ACCEPTED NAMES
Dendrobium hamatum	
Dendrobium hamiferum	
Dendrobium hanburyanum	Dendrobium lituiflorum
Dendrobium hancockii	
Dendrobium haniffii	Dendrobium tortile
Dendrobium haniffii var. *dartoisianum*	Dendrobium tortile
Dendrobium harveyanum	
Dendrobium hasseltii*	
Dendrobium hasseltii	Dendrobium malvicolor
Dendrobium hawkesii	Dendrobium fimbriatum
Dendrobium hedyosmum	Dendrobium scabrilingue
Dendrobium helenae	Dendrobium rigidifolium
Dendrobium helix	
Dendrobium hellwigianum	
Dendrobium hemimelanoglossum	
Dendrobium hendersonii	
Dendrobium henryi	
Dendrobium herbaceum	
Dendrobium hercoglossum*	
Dendrobium hercoglossum	Dendrobium faulhaberianum
Dendrobium hercoglossum	Dendrobium linguella
Dendrobium heterocarpum	
Dendrobium heterocaulon	Dendrobium exile
Dendrobium heterostigma	Dendrobium secundum
Dendrobium heyneanum	
Dendrobium hildebrandii	Dendrobium heterocarpum*
Dendrobium hildebrandii	Dendrobium signatum
Dendrobium hillii	Dendrobium tarberi
Dendrobium hodgkinsonii	
Dendrobium hollrungii	Dendrobium smillieae
Dendrobium hollrungii var. *australiense*	Dendrobium smillieae
Dendrobium holttumianum	Dendrobium cinnabarinum var. angustitepalum
Dendrobium hookerianum	
Dendrobium hornei	
Dendrobium humile	Dendrobium microbolbon
Dendrobium huoshanense	
Dendrobium hymenanthum	Dendrobium hymenopterum
Dendrobium hymenocentrum	
Dendrobium hymenophyllum	
Dendrobium hymenopterum	
Dendrobium hymenopterum	Dendrobium intricatum
Dendrobium igneonivium	
Dendrobium igneoviolaceum	Dendrobium violaceum subsp. cyperifolium
Dendrobium imperatrix	Dendrobium lineale
Dendrobium imthurnii	Dendrobium gouldii
Dendrobium inamoenum	
Dendrobium inconcinnum	Dendrobium podagraria
Dendrobium incrassatum	Dendrobium indivisum
Dendrobium incurvilabium	Dendrobium dendrocolloides
Dendrobium incurvociliatum	
Dendrobium incurvum	
Dendrobium indivisum	
Dendrobium indivisum var. indivisum	
Dendrobium indivisum var. *lampangense*	Dendrobium porphyrophyllum

*For explanation see page 4, point 6
*Voir les explications page 9, point 6
*Para mayor explicación, véase la página 15, point 6

ALL NAMES	ACCEPTED NAMES
Dendrobium indivisum var. pallidum	
Dendrobium inflatum	
Dendrobium informe	
Dendrobium infractum	
Dendrobium infundibulum	
Dendrobium infundibulum var. jamesianum	Dendrobium infundibulum
Dendrobium inopinatum	Dendrobium erosum
Dendrobium × intermedium	Dendrobium × vonpaulsenianum
Dendrobium intricatum	
Dendrobium intricatum	Dendrobium cumulatum*
Dendrobium ionoglossum	Dendrobium nindii
Dendrobium ionoglossum var. potamophilum	Dendrobium nindii
Dendrobium ionopus	Dendrobium panduriferum
Dendrobium irayense	Dendrobium goldschmidtianum
Dendrobium jabiense	
Dendrobium jacobsonii	
Dendrobium jamesianum	Dendrobium infundibulum
Dendrobium japonicum	Dendrobium moniliforme
Dendrobium jenkinsii	
Dendrobium jenkinsii	Dendrobium lindleyi*
Dendrobium johannis	
Callista johannis	Dendrobium johannis
Dendrobium johannis var. semifuscum	Dendrobium trilamellatum
Dendrobium johnsoniae	
Dendrobium jonesii	
Dendrobium jonesii subsp. bancroftianium	
Dendrobium jonesii subsp. blackburnii	
Dendrobium jonesii subsp. jonesii	
Dendrobium junceum	
Dendrobium juncoideum	
Dendrobium junzaingense	Dendrobium subacaule
Dendrobium kaernbachii	Dendrobium smillieae
Dendrobium kajewskii	Dendrobium conanthum
Dendrobium kanburiense	
Dendrobium kauldorumii	
Dendrobium keithii	
Dendrobium kennedyi	Dendrobium sylvanum
Dendrobium kerewense	Dendrobium dekockii
Dendrobium × kestevenii	
Dendrobium kestevenii var. coloratum	Dendrobium × kestevenii
Dendrobium keysseri	Dendrobium nebularum
Dendrobium keytsianum	
Dendrobium kiauense	
Dendrobium kingianum	
Dendrobium kingianum var. aldersoniae	Dendrobium kingianum
Dendrobium kingianum var. kestevenii	Dendrobium × kestevenii
Dendrobium kingianum var. pallidum	Dendrobium kingianum
Dendrobium kingianum var. pulcherrimum	Dendrobium kingianum
Dendrobium kingianum var. silcockii	Dendrobium kingianum
Dendrobium kingianum var. suffusum	Dendrobium × suffusum
Dendrobium klabatense	
Dendrobium koeteianum	Dendrobium aciculare
Dendrobium kontumense	Dendrobium virgineum
Dendrobium korthalsii	
Dendrobium kraemeri	
Dendrobium kraemeri var. pseudokraemeri	Dendrobium kraemeri

*For explanation see page 4, point 6
*Voir les explications page 9, point 6
*Para mayor explicación, véase la página 15, point 6

ALL NAMES	ACCEPTED NAMES
Dendrobium kratense	
Dendrobium kruiense	
Dendrobium kuhlii	Dendrobium hasseltii
Dendrobium kuhlii	Dendrobium malvicolor
Dendrobium kuhlii	Dendrobium thyrsodes
Dendrobium kwashotense	Dendrobium crumenatum
Dendrobium laetum	Dendrobium cuthbertsonii
Dendrobium laevifolium	
Dendrobium lamellatum	
Dendrobium lamelluliferum	
Dendrobium lamii	
Dendrobium lampongense	
Dendrobium lancifolium	
Dendrobium lancilabium	
Dendrobium lancilobum	
Dendrobium lanepoolei	
Dendrobium langbianense	
Dendrobium lanyaiae	
Dendrobium lapeyrouseiodes	Dendrobium cyanocentrum
Dendrobium lasianthera	
Dendrobium lasioglossum	
Dendrobium lateriflorum	Dendrobium puniceum
Dendrobium laurensii	
Dendrobium lauterbachianum	Dendrobium obtusum
Dendrobium × lavarackianum	
Dendrobium lawanum	Dendrobium crepidatum
Dendrobium lawesii	
Dendrobium lawiense	
Dendrobium laxiflorum	
Dendrobium leonis	
Dendrobium leonis var. *strictum*	Dendrobium terminale
Dendrobium leporinum	
Dendrobium leucochlorum	
Dendrobium leucochysum	Dendrobium bracteosum
Dendrobium leucocyanum	
Dendrobium leucohybos	
Dendrobium leucohybos var. *leucanthum*	Dendrobium leucohybos
Dendrobium leucolophotum	Dendrobium affine
Dendrobium leucorhodum	Dendrobium anosmum
Dendrobium lichenastrum	
Dendrobium lichenastrum var. *prenticei*	Dendrobium prenticei
Dendrobium lichenicola	Dendrobium cuthbertsonii
Dendrobium lilacinum	Dendrobium swartzii
Dendrobium limii	
Dendrobium linawianum	
Dendrobium lindleyanum	Dendrobium nobile
Dendrobium lindleyi	
Dendrobium lineale	
Dendrobium linearifolium	
Dendrobium linguella	
Dendrobium linguella	Dendrobium hercoglossum*
Dendrobium linguiforme	
Dendrobium linguiforme var. *nugentii*	Dendrobium nugentii
Dendrobium listeroglossum	Dendrobium parcum
Dendrobium lithocola	
Dendrobium litorale	

*For explanation see page 4, point 6
*Voir les explications page 9, point 6
*Para mayor explicación, véase la página 15, point 6

Part I: All Names / Tous les Noms / Todos los Nombres

ALL NAMES	ACCEPTED NAMES
Dendrobium lituiflora	Dendrobium lituiflorum
Dendrobium lituiflorum	
Dendrobium lobatum	
Dendrobium lobulatum	
Dendrobium loddigesii	
Dendrobium loesenerianum	
Dendrobium lohoense	
Dendrobium lomatochilum	
Dendrobium lompobatangense	Dendrobium caliculimentum
Dendrobium longicalcaratum	Dendrobium chameleon
Dendrobium longicornu*	
Dendrobium longicornu	Dendrobium wattii
Dendrobium longicornu	Dendrobium williamsonii
Dendrobium lowii	
Dendrobium lowii var. *pleiotrichum*	Dendrobium lowii
Dendrobium lubbersianum	Dendrobium williamsonii
Dendrobium lucens	
Dendrobium lueckelianum	
Dendrobium lunatum	
Dendrobium maboroense	Dendrobium undatialatum
Dendrobium macarthiae	
Dendrobium macfarlanei	Dendrobium eboracense
Dendrobium macfarlanei	Dendrobium johnsoniae*
Dendrobium macgregorii	
Dendrobium macranthum	
Dendrobium macranthum	Dendrobium anosmum*
Dendrobium macrifolium	
Dendrobium macrogenion	
Dendrobium macrophyllum	
Dendrobium macrophyllum	Dendrobium anosmum*
Dendrobium macrophyllum var. **macrophyllum**	
Dendrobium macrophyllum var. *stenopterum*	Dendrobium polysema
Dendrobium macrophyllum var. **subvelutinum**	
Dendrobium macrophyllum var. *veitchianum*	Dendrobium macrophyllum
Dendrobium macropus	
Dendrobium macropus subsp. *howeanum*	Dendrobium comptonii
Dendrobium macrostachyum	
Dendrobium madonnae	Dendrobium rhodostictum
Dendrobium madrasense	Dendrobium aphyllum
Dendrobium magistratus	
Dendrobium magnificum	Dendrobium terrestre
Dendrobium maierae	
Dendrobium malvicolor	
Dendrobium mannii	
Dendrobium mannii	Dendrobium nathanielis
Dendrobium margaritaceum	Dendrobium christyanum
Dendrobium marginatum	Dendrobium xanthophlebium
Dendrobium maritimum	Dendrobium fulgidum var. **maritimum**
Dendrobium marmoratum	
Dendrobium marseilleii	Dendrobium jenkinsii
Dendrobium masarangense	
Dendrobium masarangense subsp. **chlorinum**	
Dendrobium masarangense subsp. **masarangense**	
Dendrobium masarangense var. **theionanthum**	
Dendrobium mayandyi	
Dendrobium melaleucaphilum	

*For explanation see page 4, point 6
*Voir les explications page 9, point 6
*Para mayor explicación, véase la página 15, point 6

44

ALL NAMES	ACCEPTED NAMES
Dendrobium melanolasium	Dendrobium-finisterrae
Dendrobium melanophthalmum	Dendrobium pendulum
Dendrobium melinanthum	
Dendrobium mellitum	Dendrobium clavator
Dendrobium micranthum	Dendrobium aloifolium
Dendrobium microblepharum	Dendrobium vexillarius var. microblepharum
Dendrobium microbolbon	
Dendrobium microbolbon	Dendrobium peguanum
Dendrobium militare	
Dendrobium milliganii	Dendrobium striolatum
Dendrobium mimiense	Dendrobium constrictum
Dendrobium minahassae	Dendrobium heterocarpum
Dendrobium minax	Dendrobium bicaudatum
Dendrobium minimum	
Dendrobium minutiflorum	
Dendrobium minutum	Dendrobium delicatulum
Dendrobium mirbelianum	
Dendrobium mitriferum	Dendrobium subclausum
Dendrobium miyakei	
Dendrobium modestissimum	
Dendrobium modestum	
Dendrobium mohlianum	
Dendrobium molle	
Dendrobium monile	Dendrobium moniliforme
Dendrobium moniliforme	
Dendrobium monodon	Dendrobium johnsoniae
Dendrobium monogrammoides	Dendrobium masarangense var. theionanthum
Dendrobium monophyllum	
Dendrobium montanum	
Dendrobium monticola	
Dendrobium montigenum	Dendrobium dekockii
Dendrobium montis-yulei	
Dendrobium montistellare	Dendrobium brevicaule subsp. calcarium
Dendrobium mooreanum	
Dendrobium moorei	
Dendrobium mortii	
Dendrobium mortii	Dendrobium schoeninum*
Dendrobium moschatum	
Dendrobium moseleyi	Dendrobium purpureum
Dendrobium moulmeinense	
Dendrobium mucronatum	
Dendrobium multiflorum	Dendrobium nathanielis
Dendrobium multiflorum	Dendrobium sarawakense
Dendrobium multilineatum	
Dendrobium multiramosum	
Dendrobium murkelense	Dendrobium nebularum
Dendrobium musciferum	Dendrobium macrophyllum
Dendrobium mutabile	
Dendrobium mystroglossum	
Dendrobium nanarauticola	Dendrobium delicatulum
Dendrobium nardoides	
Dendrobium nathanielis	
Dendrobium nathanielis	Dendrobium mannii

*For explanation see page 4, point 6
*Voir les explications page 9, point 6
*Para mayor explicación, véase la página 15, point 6

ALL NAMES	ACCEPTED NAMES
Dendrobium navicula	
Dendrobium nebularum	
Dendrobium neo-ebudanum	Dendrobium mohlianum
Dendrobium neolampangense	Dendrobium porphyrophyllum
Dendrobium nieuwenhuisii	
Dendrobium nindii	
Dendrobium niveum	Dendrobium johnsoniae
Dendrobium nobile	
Dendrobium nobile var. alboluteum	
Dendrobium nobile var. formosanum	Dendrobium nobile
Dendrobium nobile var. nobile	
Dendrobium nobile var. pallidiflora	Dendrobium primulinum
Dendrobium nothofagicola	
Dendrobium novae-hiberniae	Dendrobium bracteosum
Dendrobium nubigenum	
Dendrobium nudum	
Dendrobium nugentii	
Dendrobium nycteridoglossum	
Dendrobium obcordatum	
Dendrobium obcuneatum	Dendrobium nugentii
Dendrobium obrienianum	
Dendrobium obtusisepalum	Dendrobium subclausum var. speciosum
Dendrobium obtusum	
Dendrobium occultum	Dendrobium laevifolium
Dendrobium ochraceum	
Dendrobium ochreatum	
Dendrobium odoardii	
Dendrobium officinale	
Dendrobium okinawense	
Dendrobium oliganthum	
Dendrobium oligoblepharon	Dendrobium nardoides
Dendrobium oligophyllum	
Dendrobium ophioglossum	Dendrobium smillieae*
Dendrobium ophioglossum	Dendrobium capituliflorum
Dendrobium oreocharis	Dendrobium subacaule
Dendrobium oreodoxa	
Dendrobium oreogenum	
Dendrobium oscari	Dendrobium macropus
Dendrobium ostrinoglossum	Dendrobium lasianthera
Dendrobium otaguroanum	
Dendrobium ovatum	
Dendrobium ovipostoriferum	
Dendrobium oxyanthum	Dendrobium faulhaberianum
Dendrobium oxyphyllum	Dendrobium aphyllum
Dendrobium paathii	
Dendrobium pachyceras	Dendrobium smillieae
Dendrobium pachyglossum	
Dendrobium pachystele	
Dendrobium pachystele var. homeoglossum	Dendrobium pachystele
Dendrobium palmerstoniae	Dendrobium adae
Dendrobium palpebrae	
Dendrobium palustre	Dendrobium nebularum
Dendrobium panduriferum	
Dendrobium panduriferum var. serpens	Dendrobium panduriferum
Dendrobium paniferum	
Dendrobium papilio	

*For explanation see page 4, point 6
*Voir les explications page 9, point 6
*Para mayor explicación, véase la página 15, point 6

46

ALL NAMES	ACCEPTED NAMES
Dendrobium papiloniferum	
Dendrobium papiloniferum var. ephemerum	
	Dendrobium papiloniferum var. ephemerum
Dendrobium papiloniferum var. papiloniferum	
Dendrobium papuanum	
Dendrobium parciflorum	
Dendrobium parcoides	Dendrobium parcum
Dendrobium parcum	
Dendrobium parishii	
Dendrobium parthenium	
Dendrobium parvifolium	
Dendrobium parvulum	Dendrobium delicatulum subsp. parvulum
Dendrobium paspalifolium	
Dendrobium patentilobum	
Dendrobium pauciflorum	
Dendrobium paxtonii	Dendrobium chrysanthum
Dendrobium paxtonii	Dendrobium fimbriatum*
Dendrobium peculiare	Dendrobium setifolium
Dendrobium pedicellatum	
Dendrobium pedunculatum	
Dendrobium peguanum	
Dendrobium pendulum	
Dendrobium pentagonum	Dendrobium brevicaule subsp. pentagonum
Dendrobium pentagonum	Dendrobium vexillarius var. retroflexum*
Dendrobium pentapterum	
Dendrobium percnanthum	
Dendrobium pere-fauriei	Dendrobium tosaense
Dendrobium perulatum	
Dendrobium petiolatum	
Dendrobium petri	Dendrobium mooreanum
Dendrobium phalaenopsis	
Dendrobium phalaenopsis var. *compactum*	Dendrobium lithocola
Dendrobium phlox	Dendrobium subclausum
Dendrobium pictum	
Dendrobium pictum	Dendrobium devonianum*
Dendrobium pierardii	Dendrobium aphyllum
Dendrobium pityphyllum	Dendrobium violaceum
Dendrobium planibulbe	
Dendrobium platyphyllum	Dendrobium nycteridoglossum
Dendrobium pleurodes	
Dendrobium podagraria	
Dendrobium pogoniates	
Dendrobium poilanei	Dendrobium hercoglossum*
Dendrobium poilanei	Dendrobium linguella
Dendrobium polyanthum	Dendrobium cretaceum
Dendrobium polycarpum	Dendrobium mirbelianum
Dendrobium polyphlebium	Dendrobium rhodopteryguim
Dendrobium polysema	
Dendrobium polysema var. *pallidum*	Dendrobium polysema
Dendrobium porphyrochilum	
Dendrobium porphyrophyllum	
Dendrobium praetermissum	

*For explanation see page 4, point 6
*Voir les explications page 9, point 6
*Para mayor explicación, véase la página 15, point 6

ALL NAMES	ACCEPTED NAMES
Dendrobium praeustum	Dendrobium purpureum
Dendrobium prasinum	
Dendrobium prenticei	
Dendrobium prianganense	
Dendrobium primulinum	
Dendrobium prionochilum	Dendrobium sylvanum
Dendrobium priscillae	Dendrobium mooreanum
Dendrobium pristinum	
Dendrobium profusum	
Dendrobium prostratum	
Dendrobium prostratum	Dendrobium intricatum
Dendrobium proteranthum	
Dendrobium pseudo-hainanense	Dendrobium miyakei
Dendrobium pseudo-kraemeri	Dendrobium kraemeri
Dendrobium pseudoaloifolium	
Dendrobium pseudocalceolum	
Dendrobium pseudoconanthum	
Dendrobium pseudofrigidum	Dendrobium masarangense var. theionanthum
Dendrobium pseudoglomeratum	
Dendrobium pseudointricatum	
Dendrobium pseudomohlianum	Dendrobium lawesii
Dendrobium pseudopeloricum	
Dendrobium pseudorarum	
Dendrobium pseudorarum var. baciforme	
Dendrobium pseudorarum var. pseudorarum	
Dendrobium pseudotenellum	
Dendrobium pseudotokai	Dendrobium macranthum
Dendrobium psyche	Dendrobium macrophyllum
Dendrobium puberilingue	
Dendrobium pugioniforme	
Dendrobium pulchellum*	
Dendrobium pulchellum	Dendrobium anosmum
Dendrobium pulchellum var. *devonianum*	Dendrobium devonianum
Dendrobium pulchrum	Dendrobium polysema
Dendrobium pumilio	Dendrobium masarangense
Dendrobium punamense	
Dendrobium pugentifolium	Dendrobium pugioniforme
Dendrobium puniceum	
Dendrobium purpureiflorum	
Dendrobium purpureum	Dendrobium catillare
Dendrobium purpureum	Dendrobium purpureum var. purpureum*
Dendrobium purpureum subsp. candidulum	
Dendrobium purpureum var. *album*	Dendrobium purpureum
Dendrobium purpureum var. *candidulum*	Dendrobium purpureum subsp. candidulum
Dendrobium purpureum var. *moseleyi*	Dendrobium purpureum
Dendrobium purpureum subsp. purpureum	
Dendrobium purpureum var. *steffensianum*	Dendrobium purpureum subsp. candidulum
Dendrobium pusillum	Dendrobium monticola
Dendrobium putnamii	
Dendrobium pychnostachyum	
Dendrobium pygmaeum	Dendrobium peguanum
Dendrobium pyropum	Dendrobium crocatum

*For explanation see page 4, point 6
*Voir les explications page 9, point 6
*Para mayor explicación, véase la página 15, point 6

ALL NAMES

Dendrobium quadrialatum
Dendrobium quadriquetrum
Dendrobium quaifei
Dendrobium quinquecostatum
Dendrobium quinquecristatum
Dendrobium racemosum
Dendrobium rachmatii
Dendrobium radians
Dendrobium rajanum
Dendrobium ramosissimum
Dendrobium ramosii
Dendrobium randaiense
Dendrobium rantii
Dendrobium rappardii
Dendrobium rariflorum
Dendrobium rarum
Dendrobium rarum
Dendrobium rarum

Dendrobium rarum var. miscegeneum
Dendrobium rarum var. pelorium
Dendrobium rarum var. rarum
Dendrobium reflexibarbatulum
Dendrobium reflexitepalum
Dendrobium reinwardtii
Dendrobium reinwardtii
Dendrobium rennellii
Dendrobium reticulatum
Dendrobium retroflexum

Dendrobium rex
Dendrobium rhabdoglossum
Dendrobium rhaphiotes
Dendrobium rhizophoreti
Dendrobium rhodobotrys
Dendrobium rhododiodes
Dendrobium rhodopterygium
Dendrobium rhodostele
Dendrobium rhodostictum
Dendrobium rhombeum
Dendrobium rhomboglossum
Dendrobium ridleyanum
Dendrobium rigidifolium
Dendrobium rigidum
Dendrobium rimannii
Dendrobium rindjaniense
Dendrobium riparium
Dendrobium rivesii
Dendrobium robertsii
Dendrobium robustum
Dendrobium rolfei
Dendrobium roseicolor
Dendrobium roseipes
Dendrobium rosellum
Dendrobium rosenbergii
Dendrobium roseum

ACCEPTED NAMES
Dendrobium bracteosum

Dendrobium mooreanum
Dendrobium violaceum
Dendrobium brevicaule var. calcarium

Dendrobium kiauense
Dendrobium herbaceum

Dendrobium chameleon

Dendrobium pseudorarum
Dendrobium pseudorarum var.
    baciforme

Dendrobium bullenianum*
Dendrobium purpureum

Dendrobium spectatissimum
Dendrobium vexillarius var.
    retroflexum

Dendrobium hellwigianum
Dendrobium lobatum
Dendrobium obtusum
Dendrobium caliculimentum

Dendrobium heterocarpum

Dendrobium hendersonii

Dendrobium calophyllum

Dendrobium chryseum
Dendrobium mortii
Dendrobium sylvanum
Dendrobium chryseum

Dendrobium mirbelianum
Dendrobium roseicolor

*For explanation see page 4, point 6
*Voir les explications page 9, point 6
*Para mayor explicación, véase la página 15, point 6

ALL NAMES	ACCEPTED NAMES
Dendrobium roseum	**Dendrobium crepidatum***
Dendrobium roxburghii	**Dendrobium calceolum**
Dendrobium roylei	**Dendrobium monticola**
Dendrobium rudolphii	**Dendrobium hendersonii**
Dendrobium ruginosum	
Dendrobium ruidilobum	**Dendrobium cochliodes**
Dendrobium rumphianum	**Dendrobium bicaudatum**
Dendrobium rupestre	
Dendrobium rupicola	**Dendrobium venustum**
Dendrobium rupicola var. *breve*	**Dendrobium delacourii**
Dendrobium ruppianum	**Dendrobium jonesii**
Dendrobium ruppianum var. *blackburnii*	**Dendrobium jonesii** subsp. **blackburnii**
Dendrobium × ruppiosum	
Dendrobium rutriferum	
Dendrobium ruttenii	
Dendrobium sacculiferum	**Dendrobium dillonianum**
Dendrobium sagittatum	
Dendrobium salmoneum	
Dendrobium salmonicolor	**Dendrobium lawesii**
Dendrobium sambasanum	
Dendrobium samoense	
Dendrobium sancristobalense	
Dendrobium sanderae	
Dendrobium sanderianum	
Dendrobium sanguineum	**Dendrobium cinnabarinum** var. **angustitepalum**
Dendrobium sanguinolentum	
Dendrobium sanguinolentum	**Dendrobium calicopis**
Dendrobium sanguinolentum	**Dendrobium cumulatum***
Dendrobium sarawakense	
Dendrobium sarcanthum	**Dendrobium cuspidatum**
Dendrobium sarcopodioides	**Dendrobium simplex**
Dendrobium sarcostoma	**Dendrobium macrophyllum**
Dendrobium saruwagedicum	**Dendrobium brevicaule** subsp. **pentagonum**
Dendrobium sayeria	**Dendrobium cruttwellii**
Dendrobium scabrifolium	
Dendrobium scabrilingue	
Dendrobium scabripes	**Dendrobium purpureum**
Dendrobium scarlatinum	**Dendrobium puniceum**
Dendrobium schmidtianum	**Dendrobium crumenatum**
Dendrobium schneiderae	
Dendrobium schneiderae var. **major**	
Dendrobium schneiderae var. **schneiderae**	
Dendrobium schoeninum	
Dendrobium schroderi	
Dendrobium schuetzei	
Dendrobium schulleri	
Dendrobium × schumannianum	
Dendrobium scortechinii	**Dendrobium anosmum**
Dendrobium scotiiferum	**Dendrobium violaceum** var. **cyperifolium**
Dendrobium sculptum	
Dendrobium sculptum	**Dendrobium virgineum***
Dendrobium secundum*	
Dendrobium secundum	**Dendrobium catillare**

*For explanation see page 4, point 6
*Voir les explications page 9, point 6
*Para mayor explicación, véase la página 15, point 6

ALL NAMES	ACCEPTED NAMES
Dendrobium secundum var. *bursigerum*	Dendrobium secundum
Dendrobium secundum var. *niveum*	Dendrobium secundum
Dendrobium secundum var. *urvillei*	Dendrobium smillieae
Dendrobium seemannii	Dendrobium vagans
Dendrobium semeion	Dendrobium vexillarius
Dendrobium semifuscum	Dendrobium trilamellatum
Dendrobium senile	
Dendrobium separatum	Dendrobium calcaratum
Dendrobium seranicum	
Dendrobium serpens	Dendrobium panduriferum
Dendrobium serra	Dendrobium aloifolium
Dendrobium sertatum	Dendrobium catillare
Dendrobium sertatum	Dendrobium taveuniense
Dendrobium setifolium	
Dendrobium sidikalangense	
Dendrobium signatum	
Dendrobium sikinii	Dendrobium subquadratum
Dendrobium sikkimense	Dendrobium pauciflorum
Dendrobium simondii	
Dendrobium simplex	
Dendrobium simplicissimum	Dendrobium crumenatum
Dendrobium sinense	
Dendrobium singalanense	Dendrobium hymenopterum
Dendrobium singkawangense	
Dendrobium sinuosum	
Dendrobium smillieae	
Dendrobium smillieae var. *hollrungii*	Dendrobium smillieae
Dendrobium smillieae var. *ophiglossum*	Dendrobium smillieae
Dendrobium smithianum	
Dendrobium somai	
Dendrobium sophronites	Dendrobium cuthbertsonii
Dendrobium soriense	
Dendrobium spatella	Dendrobium acinaciforme
Dendrobium spathilingue	
Dendrobium spathulatilabratum	Dendrobium habbemense
Dendrobium spathulatum	
Dendrobium speciosissimum	Dendrobium spectatissimum
Dendrobium speciosum	
Dendrobium speciosum var. *album*	Dendrobium × delicatum
Dendrobium speciosum var. *bancroftianum*	Dendrobium jonesii subsp. bancroftianium
Dendrobium speciosum var. *curvicaule*	Dendrobium curvicaule
Dendrobium speciosum var. *delicatum*	Dendrobium × delicatum
Dendrobium speciosum var. *fusiforme*	Dendrobium jonesi
Dendrobium speciosum var. *gracillimum*	Dendrobium × gracillimum
Dendrobium speciosum var. *grandiflorum*	Dendrobium rex
Dendrobium speciosum var. *pedunculatum*	Dendrobium pedunculatum
Dendrobium spectabile	
Dendrobium spectatissimum	
Dendrobium spegidoglossum	
Dendrobium spegidoglossum	Dendrobium stuposum
Dendrobium stellare	
Dendrobium stolleanum	
Dendrobium stratiotes	
Dendrobium strebloceras	
Dendrobium strebloceras var. *rossianum*	Dendrobium stratiotes

*For explanation see page 4, point 6
*Voir les explications page 9, point 6
*Para mayor explicación, véase la página 15, point 6

ALL NAMES	ACCEPTED NAMES
Dendrobium strepsiceros	
Dendrobium strictum	Dendrobium subclausum
Dendrobium strigosum	Dendrobium tetraedre
Dendrobium striolatum	
Dendrobium striolatum	Dendrobium schoeninum*
Dendrobium striolatum var. *beckleri*	Dendrobium schoeninum
Dendrobium striolatum var. *chalandei*	Dendrobium bowmanii
Dendrobium strongylanthum	
Dendrobium stuartii	
Dendrobium stueberi	Dendrobium lasianthera
Dendrobium stuposum	
Dendrobium stuposum	Dendrobium spegidoglossum
Dendrobium suavissimum	Dendrobium chrysotoxum
Dendrobium suavissimum	Dendrobium lindleyi
Dendrobium subacaule*	
Dendrobium subacaule	Dendrobium puniceum
Dendrobium subclausum	
Dendrobium subclausum var. pandanicola	
Dendrobium subclausum var. phlox	
Dendrobium subclausum var. speciosum	
Dendrobium subclausum var. subclausum	
Dendrobium subquadratum	
Dendrobium subulatoides	
Dendrobium subuliferum	
Dendrobium subuliferum var. *gautierense*	Dendrobium subuliferum
Dendrobium × suffusum	
Dendrobium sulcatum	
Dendrobium sulphureum	
Dendrobium sulphureum var. cellulosum	
Dendrobium sulphureum var. rigidifolium	
Dendrobium sulphureum var. sulphureum	
Dendrobium superbiens	Dendrobium × superbiens
Dendrobium × superbiens	
Dendrobium superbum	Dendrobium anosmum
Dendrobium sutepense	
Dendrobium swartzii	
Dendrobium sylvanum	
Dendrobium takahashii	
Dendrobium talaudense	Dendrobium purpureum subsp. candidulum
Dendrobium tangerinum	
Dendrobium tapiniense	
Dendrobium tarberi	
Dendrobium tattonianum	
Dendrobium taurinum	
Dendrobium taurulinum	
Dendrobium taveuniense	
Dendrobium teligerum	Dendrobium brevicaule subsp. pentagonum
Dendrobium tenellum	
Dendrobium tenellum	Dendrobium pseudotenellum
Dendrobium tenellum var. *setifolium*	Dendrobium pseudotenellum
Dendrobium tenens	Dendrobium vexillarius var. uncinatum
Dendrobium tenue	
Dendrobium tenuicalcar	Dendrobium violaceum
Dendrobium tenuissimum	Dendrobium mortii

*For explanation see page 4, point 6
*Voir les explications page 9, point 6
*Para mayor explicación, véase la página 15, point 6

ALL NAMES	ACCEPTED NAMES
Dendrobium teretifolium	
Dendrobium teretifolium	Dendrobium striolatum*
Dendrobium teretifolium var. *album*	Dendrobium calamiforme
Dendrobium teretifolium var. *aureum*	Dendrobium dolichophyllum
Dendrobium teretifolium var. *fairfaxii*	Dendrobium fairfaxii
Dendrobium teretifolium var. *fasciculatum*	Dendrobium calamiforme
Dendrobium terminale	
Dendrobium terminale	Dendrobium mannii
Dendrobium ternatense	Dendrobium macrophyllum
Dendrobium terrestre	
Dendrobium terrestre var. *sublobatum*	Dendrobium terrestre
Dendrobium tetrachromum	
Dendrobium tetraedre	
Dendrobium tetraedre	Dendrobium exile
Dendrobium tetragonum	
Dendrobium tetragonum var. *giganteum*	Dendrobium cacatua
Dendrobium tetralobum	
Dendrobium tetrodon	
Dendrobium tetrodon	Dendrobium stuartii
Dendrobium theionanthum	Dendrobium masarangense var. theionanthum
Dendrobium thyrsiflorum	
Dendrobium thyrsodes	
Dendrobium tibeticum	Dendrobium chryseum
Dendrobium tigrinum	Dendrobium spectabile
Dendrobium tixieri	Dendrobium oligophyllum
Dendrobium tofftii	Dendrobium nindii
Dendrobium tokai	
Dendrobium tokai var. *crassinerve*	Dendrobium macranthum
Dendrobium tomohonense	Dendrobium macrophyllum
Dendrobium topaziacum	Dendrobium bullenianum
Dendrobium toressae	
Dendrobium torricellense	
Dendrobium tortile*	
Dendrobium tortile	Dendrobium monophyllum
Dendrobium tortile var. *dartoisianum*	Dendrobium tortile
Dendrobium tortile var. *hildebrandii*	Dendrobium signatum
Dendrobium tortile var. *simondii*	Dendrobium tortile
Dendrobium tosaense	
Dendrobium tosaense var. *pere-fauriei*	Dendrobium tosaense
Dendrobium tozerensis	
Dendrobium trachyphyllum	Dendrobium cuthbertsonii
Dendrobium trachythece	
Dendrobium transparens	
Dendrobium transtilliferum	
Dendrobium trialatum	Dendrobium vexillarius var. uncinatum
Dendrobium trichostomum	
Dendrobium tricostatum	Dendrobium subacaule
Dendrobium tricuspe	
Dendrobium tridentatum	
Dendrobium trifolium	Dendrobium vexillarius var. uncinatum
Dendrobium trigonopus	
Dendrobium trilamellatum	
Dendrobium trisaccatum	Dendrobium bracteosum
Dendrobium triviale	Dendrobium calcaratum
Dendrobium truncatum	

*For explanation see page 4, point 6
*Voir les explications page 9, point 6
*Para mayor explicación, véase la página 15, point 6

ALL NAMES	ACCEPTED NAMES
Dendrobium tuberiferum	Dendrobium planibulbe
Dendrobium tubiflorum	
Dendrobium tumidulum	Dendrobium nebularum
Dendrobium uliginosum	
Dendrobium umbonatum	
Dendrobium uncatum	
Dendrobium uncinatum	Dendrobium vexillarius var. uncinatum
Dendrobium uncipes	
Dendrobium undatialatum	
Dendrobium undulatum	Dendrobium discolor
Dendrobium undulatum var. *broomfieldii*	Dendrobium discolor var. broomfieldii
Dendrobium undulatum var. *fimbrilabium*	Dendrobium discolor var. fimbrilabium
Dendrobium undulatum var. *johannis*	Dendrobium johannis
Dendrobium undulatum var. *woodfordianum*	Dendrobium gouldii
Dendrobium unicum	
Dendrobium unifoliatum	Dendrobium petiolatum
Dendrobium urvillei	Dendrobium affine
Dendrobium usterioides	
Dendrobium vagans	
Dendrobium vagans	Dendrobium fimbriatum*
Dendrobium validum	Dendrobium sylvanum
Dendrobium vannouhuysii	
Dendrobium variabile	Dendrobium prenticei
Dendrobium veitchianum	Dendrobium macrophyllum
Dendrobium velutinum	Dendrobium trigonopus
Dendrobium ventricosum	
Dendrobium ventrilabium	
Dendrobium ventripes	
Dendrobium venustum	
Dendrobium veratrifolium	Dendrobium lineale
Dendrobium veratroides	Dendrobium lineale
Dendrobium verlaquii	Dendrobium terminale
Dendrobium verruculosum	
Dendrobium vexans	Dendrobium hercoglossum
Dendrobium vexillarius*	
Dendrobium vexillarius	Dendrobium vexillarius var. microblepharum
Dendrobium vexillarius var. albiviride	
Dendrobium vexillarius var. elworthyi	
Dendrobium vexillarius var. microblepharum	
Dendrobium vexillarius var. retroflexum	
Dendrobium vexillarius var. uncinatum	
Dendrobium vexillarius var. vexillarius	
Dendrobium victoriae-reginae	
Dendrobium victoriae-reginae var. *miyakei*	Dendrobium miyakei
Dendrobium × vinicolor	Dendrobium × superbiens
Dendrobium violaceoflavens	
Dendrobium violaceominiatum	
Dendrobium violaceum	
Dendrobium violaceum subsp. cyperifolium	
Dendrobium violaceum subsp. violaceum	
Dendrobium violascens	
Dendrobium virescens	Dendrobium panduriferum
Dendrobium virgineum	
Dendrobium viridiroseum	Dendrobium purpureum
Dendrobium viridiroseum var. *candidulum*	Dendrobium purpureum subsp.

*For explanation see page 4, point 6
*Voir les explications page 9, point 6
*Para mayor explicación, véase la página 15, point 6

54

ALL NAMES	ACCEPTED NAMES
	candidulum
Dendrobium viriditepalum	
Dendrobium vitellinum	Dendrobium mohlianum
Dendrobium wallichii	Dendrobium peguanum
Dendrobium wangii	Dendrobium hercoglossum
Dendrobium warburgianum	Dendrobium lawesii
Dendrobium wardianum	
Dendrobium warianum	Dendrobium sylvanum
Dendrobium wassellii	
Dendrobium waterhousei	Dendrobium punamense
Dendrobium wattii	
Dendrobium wattii	Dendrobium williamsonii*
Dendrobium wentianum	
Dendrobium wenzelii	
Dendrobium whistleri	
Dendrobium whiteanum	Dendrobium stuartii
Dendrobium wightii	Dendrobium graminifolium
Dendrobium wilkianum	Dendrobium mirbelianum
Dendrobium williamsianum	
Dendrobium williamsonii	
Dendrobium wilmsianum	
Dendrobium wilsonii	
Dendrobium wisselense	
Dendrobium wollastonii	Dendrobium eximium
Dendrobium woluense	
Dendrobium womersleyi	
Dendrobium womersleyi var. autophilum	
Dendrobium womersleyi var. womersleyi	
Dendrobium woodfordianum	Dendrobium gouldii
Dendrobium woodsii	
Dendrobium wulaiense	
Dendrobium xanthellum	Dendrobium fulgidum
Dendrobium xanthoacron	
Dendrobium xanthogenium	
Dendrobium xanthophlebium	
Dendrobium xichouense	
Dendrobium xiphiphorum	Dendrobium vexillarius var. uncinatum
Dendrobium xiphophyllum	
Dendrobium × yengiliense	
Dendrobium ypsilon	
Dendrobium zaranense	Dendrobium brevicaule subsp. pentagonum
Dendrobium zhaojuense	
Dendrobium zonatum	Dendrobium moniliforme
Diacrium bidentatum	Encyclia boothiana
Dinema polybulbon	Encyclia polybulbon
Disa aconitoides	
Disa aconitoides subsp. aconitoides	
Disa aconitoides subsp. concinna	
Disa aconitoides subsp. goetzeana	
Disa aconitoides var. dichroa	Disa dichroa
Disa adolphi-friderici	Disa ochrostachya
Disa aemula	Disa cornuta
Disa aequiloba	
Disa alticola	
Disa amblyopetala	Disa hircicornis

*For explanation see page 4, point 6
*Voir les explications page 9, point 6
*Para mayor explicación, véase la página 15, point 6

55

ALL NAMES	ACCEPTED NAMES
Disa amoena	
Disa andringitrana	
Disa aperta	
Disa arida	
Disa aristata	
Disa atricapilla	
Disa attenuata	Disa sagittalis
Disa aurantiaca	Disa ochrostachya
Disa aurata	
Disa bakeri	Disa stairsii
Disa basutorum	
Disa basutorum	Disa sankeyi
Disa begleyi	
Disa bisetosa	Disa aconitoides subsp. concinna
Disa bivalvata	
Disa bivalvata var. *atricapilla*	Disa atricapilla
Disa bodkinii	
Disa borbonica	
Disa brachyceras	
Disa brevipetala	
Disa breyeri	Disa welwitschii subsp. welwitschii
Disa buchenaviana	
Disa caffra	
Disa calophylla	Disa welwitschii subsp. welwitschii
Disa capricornis	Disa gladioliflora subsp. capricornis
Disa cardinalis	
Disa carsonii	Disa erubescens var. carsonii
Disa caulescens	
Disa cedarbergensis	
Disa celata	
Disa cephalotes	
Disa cephalotes subsp. cephalotes	
Disa cephalotes subsp. frigida	
Disa chiovendaei	Disa aconitoides subsp. goetzeana
Disa chrysostachya	
Disa chrysostachya	Disa polygonoides
Disa clavicornis	
Disa coccinea	Disa robusta
Disa cochlearis	
Disa compta	Disa caffra
Disa concinna	Disa aconitoides subsp. concinna
Disa concinna var. *dichroa*	Disa dichroa
Disa cooperi	
Disa cooperi var. *scullyi*	Disa scullyi
Disa cornuta	
Disa cornuta var. *aemula*	Disa cornuta
Disa crassicornis	
Disa cryptantha	
Disa culveri	Disa hircicornis
Disa cylindrica	
Disa danielae	
Disa deckenii	Disa fragrans subsp. deckenii
Disa dichroa	
Disa dracomontana	
Disa draconis	
Disa elegans	

*For explanation see page 4, point 6
*Voir les explications page 9, point 6
*Para mayor explicación, véase la página 15, point 6

56

ALL NAMES	ACCEPTED NAMES
Disa eminii	
Disa englerana	
Disa englerana	Disa ukingensis
Disa equestris	
Disa equestris var. *concinna*	Disa aconitoides subsp. concinna
Disa equestris var. *concinna*	Disa aperta
Disa erubescens	
Disa erubescens subsp. **carsonii**	
Disa erubescens subsp. **erubescens**	
Disa erubescens var. *carsonii*	Disa erubescens subsp. carsonii
Disa erubescens var. *katangensis*	Disa katangensis
Disa erubescens var. *leucantha*	Disa erubescens subsp. erubescens
Disa esterhuyseniae	
Disa excelsa	Disa tripetaloides
Disa extinctoria	
Disa falcata	Disa tripetaloides
Disa fallax	Disa incarnata
Disa fanniniae	Disa nervosa
Disa fasciata	
Disa ferruginea	
Disa filicornis	
Disa filicornis var. *latipetala*	Disa filicornis
Disa fragrans	
Disa fragrans subsp. **deckenii**	
Disa fragrans subsp. **fragrans**	
Disa frigida	Disa cephalotes subsp. frigida
Disa galpinii	
Disa gerrardii	Disa patula var. transvaalensis
Disa gladioliflora	
Disa gladioliflora	Disa gladioliflora subsp. capricornis
Disa gladioliflora subsp. **capricornis**	
Disa gladioliflora subsp. **gladioliflora**	
Disa glandulosa	
Disa goetzeana	Disa aconitoides subsp. goetzeana
Disa gracilis	Disa chrysostachya
Disa grandiflora	Disa uniflora
Disa gregorana	Disa stairsii
Disa hallackii	
Disa harveiana*	
Disa harveiana	Disa draconis
Disa harveiana subsp. **harveiana**	
Disa harveiana subsp. **longicalcarata**	
Disa hemispaerophora	Disa versicolor
Disa hircicornis	
Disa huillensis	Disa equestris
Disa huttonii	Disa sanguinea
Disa hyacinthina	Disa welwitschii subsp. welwitschii
Disa ignea	Disa welwitschii subsp. welwitschii
Disa incarnata	
Disa intermedia	
Disa introrsa	
Disa jacottetetiae	Disa crassicornis
Disa karooica	
Disa katangensis	
Disa katangensis var. *katangensis*	Disa katangensis
Disa kilimanjarica	Disa fragrans var. deckenii

*For explanation see page 4, point 6
*Voir les explications page 9, point 6
*Para mayor explicación, véase la página 15, point 6

ALL NAMES	ACCEPTED NAMES
Disa kraussii	**Disa pulchra**
Disa laeta	**Disa hircicornis**
Disa leopoldii	**Disa walleri**
Disa leptostachys	**Disa tenuis**
Disa leucostachys	**Disa fragrans** subsp. **fragrans**
Disa lineata	
Disa longicornu	
Disa longifolia	
Disa lutea	**Disa tenuifolia**
Disa luxurians	**Disa stairsii**
Disa macowanii	**Disa versicolor**
Disa macrantha	**Disa cornuta**
Disa maculata*	
Disa maculata	**Disa ocellata**
Disa maculomarronina	
Disa marlothii	
Disa megaceras	**Disa crassicornis**
Disa melaleuca	**Disa bivalvata**
Disa micropetala	
Disa minax	**Disa galpinii**
Disa miniata	
Disa minor	
Disa modesta	**Disa vaginata**
Disa montana	
Disa natalensis	**Disa polygonoides**
Disa neglecta*	
Disa neglecta	**Disa lineata**
Disa nervosa	
Disa nervosa	**Disa patula**
Disa nigerica	
Disa nivea	
Disa nyassana	**Disa zombica**
Disa nyikensis	
Disa obtusa	
Disa obtusa subsp. **hottentotica**	
Disa obtusa subsp. **obtusa**	
Disa obtusa subsp. **picta**	
Disa occultans	**Disa welwitschii** var. **occultans**
Disa ocellata	
Disa ochrostachya	
Disa ochrostachya var. *latipetala*	**Disa satyriopsis**
Disa oligantha	
Disa oreophila	
Disa oreophila subsp. **erecta**	
Disa oreophila subsp. **oreophila**	
Disa ornithantha	
Disa ovalifolia	
Disa pappei	**Disa obtusa** subsp. **picta**
Disa parvilabris	**Disa oligantha**
Disa patens	**Disa filicornis**
Disa patens	**Disa tenuifolia***
Disa patula	
Disa patula var. **patula**	
Disa patula var. **transvaalensis**	
Disa perplexa	
Disa perrieri	**Disa caffra**

*For explanation see page 4, point 6
*Voir les explications page 9, point 6
*Para mayor explicación, véase la página 15, point 6

ALL NAMES	ACCEPTED NAMES
Disa picta	Disa obtusa var. picta
Disa pillansii	
Disa poikilantha	Disa montana
Disa polygonoides	
Disa porrecta	
Disa praestans	Disa robusta
Disa princeae	Disa walleri
Disa pulchella	
Disa pulchra	
Disa pulchra var. *montana*	Disa montana
Disa racemosa	
Disa racemosa var. *isopetala*	Disa racemosa
Disa racemosa var. *venosa*	Disa venosa
Disa reflexa	Disa filicornis
Disa rhodantha	
Disa richardiana	
Disa robusta	
Disa roeperocharoides	
Disa rosea	
Disa rungweensis	
Disa rungweensis subsp. *rhodesiaca*	Disa zimbabweensis
Disa rungweensis var. *rhodesiaca*	Disa zimbabweensis
Disa sagittalis	
Disa sagittalis var. *triloba*	Disa triloba
Disa salteri	
Disa sanguinea	
Disa sankeyi	
Disa satyriopsis	
Disa saxicola	
Disa schimperi	Disa scutellifera
Disa schizodioides	
Disa scullyi	
Disa scutellifera	
Disa secunda	Disa racemosa
Disa similis	
Disa stachyoides	
Disa stairsii	
Disa stenoglossa	Disa patula
Disa stokoei	Disa hallackii
Disa stolonifera	Disa eminii
Disa stolzii	Disa erubescens subsp. carsonii
Disa stricta	
Disa subaequalis	Disa welwitschii var. occultans
Disa subscutellifera	Disa englerana
Disa subtenuicornis	
Disa tabularis	Disa obtusa subsp. obtusa
Disa tanganyikensis	Disa welwitschii subsp. occultans
Disa telipogonis	
Disa tenella	
Disa tenella subsp. pusilla	
Disa tenella subsp. tenella	
Disa tenella var. *brachyceras*	Disa brachyceras
Disa tenuicornis	
Disa tenuifolia	
Disa tenuis	
Disa thodei	

*For explanation see page 4, point 6
*Voir les explications page 9, point 6
*Para mayor explicación, véase la página 15, point 6

ALL NAMES	ACCEPTED NAMES
Disa triloba	
Disa tripetaloides	
Disa tripetaloides subsp. *aurata*	Disa aurata
Disa tripetaloides var. *aurata*	Disa aurata
Disa tysonii	
Disa ukingensis	
Disa uliginosa	Disa saxicola
Disa uncinata	
Disa uniflora	
Disa vaginata	
Disa vaginata	Disa aconitoides var. goetzeana
Disa vasselotii	
Disa venosa	
Disa venosa	Disa tripetaloides*
Disa verdickii	
Disa versicolor	
Disa walleri	
Disa welwitschii	
Disa welwitschii	Disa roeperocharoides
Disa welwitschii subsp. **occultans**	
Disa welwitschii subsp. **welwitschii**	
Disa welwitschii var. *buchneri*	Disa welwitschii subsp. welwitschii
Disa wissmannii	Disa stairsii
Disa woodii	
Disa zeyheri	Disa porrecta
Disa zimbabweensis	
Disa zombaensis	Disa walleri
Disa zombica	
Disa zuluensis	
Dockrillia baseyana	Dendrobium calamiforme
Dockrillia beckleri	Dendrobium schoeninum
Dockrillia bowmanii	Dendrobium bowmanii
Dockrillia calamiformis	Dendrobium calamiforme
Dockrillia chordiformis	Dendrobium chordiforme
Dockrillia cucumerina	Dendrobium cucumerinum
Dockrillia desmotrichoides	Dendrobium rigidum
Dockrillia dolichophylla	Dendrobium dolichophyllum
Dockrillia fairfaxii	Dendrobium fairfaxii
Dockrillia flagella	Dendrobium flagellum
Dockrillia × *foederata*	Dendrobium × foederatum
Dockrillia × *grimesii*	Dendrobium × grimesii
Dockrillia lichenastrum	Dendrobium lichenastrum
Dockrillia linguiforme	Dendrobium linguiforme
Dockrillia mortii	Dendrobium mortii
Dockrillia nugentii	Dendrobium nugentii
Dockrillia pugioniformis	Dendrobium pugioniforme
Dockrillia racemosa	Dendrobium racemosum
Dockrillia rigida	Dendrobium rigidum
Dockrillia schoenina	Dendrobium schoeninum
Dockrillia striolata	Dendrobium striolatum
Dockrillia tenuissima	Dendrobium mortii
Dockrillia teretifolia	Dendrobium teretifolium
Dockrillia toressae	Dendrobium toressae
Dockrillia vagans	Dendrobium vagans
Dockrillia wassellii	Dendrobium wassellii
Domingoa kienastii	Encyclia kienastii

*For explanation see page 4, point 6
*Voir les explications page 9, point 6
*Para mayor explicación, véase la página 15, point 6

ALL NAMES	ACCEPTED NAMES
Doritis bifalcis	Dendrobium bifalce
Dracula alcithoe	
Dracula amaliae	
Dracula andreettae	
Dracula anicula	
Dracula anthracina	
Dracula aphrodes	
Dracula astuta	Dracula erythrochaete subsp. astuta
Dracula bella	
Dracula bellerophon	
Dracula benedictii	
Dracula berthae	
Dracula brangeri	
Dracula burbidgeana	Dracula erythrochaete
Dracula callifera	Dracula houtteana
Dracula carcinopsis	
Dracula carderi	Dracula inaequalis
Dracula carderiopsis	Dracula houtteana
Dracula carlueri	
Dracula chestertonii	
Dracula chimaera	
Dracula chiroptera	
Dracula × circe	
Dracula citrina	
Dracula cochliops	
Dracula cordobae	
Dracula cutis-bufonis	
Dracula dalessandroi	
Dracula dalstroemii	
Dracula decussata	
Dracula deltoidea	
Dracula diabola	
Dracula diana	
Dracula dodsonii	
Dracula erythrochaete	
Dracula erythrochaete subsp. santa-elenae	
Dracula exasperata	
Dracula fafnir	
Dracula felix	
Dracula fuligifera	
Dracula fuliginosa	Dracula radiella*
Dracula gaskelliana	Dracula erythrochaete
Dracula gastrophora	
Dracula gigas	
Dracula gorgo	Dracula erythrochaete subsp. astuta
Dracula gorgona	
Dracula gorgonella	
Dracula hawleyi	
Dracula hirsuta	
Dracula hirtzii	
Dracula houtteana	
Dracula hubeinii	Dracula benedictii
Dracula inaequalis	
Dracula incognita	
Dracula insolita	
Dracula iricolor	Dracula trichroma

*For explanation see page 4, point 6
*Voir les explications page 9, point 6
*Para mayor explicación, véase la página 15, point 6

ALL NAMES	ACCEPTED NAMES
Dracula janetiae	
Dracula lactea	Dracula velutina
Dracula lafleurii	
Dracula lehmanniana	
Dracula lemurella	
Dracula leonum	Dracula erythrochaete
Dracula levii	
Dracula ligiae	
Dracula lindstroemii	
Dracula lotax	
Dracula lowii	Dracula platycrater
Dracula mantissa	
Dracula marsupialis	
Dracula medellinensis	Dracula radiosa
Dracula microglochin	Dracula velutina
Dracula minax	
Dracula mopsus	
Dracula morleyi	
Dracula mosquerae	Dracula houtteana
Dracula navarroorum	
Dracula niesseniae	Dracula descussata
Dracula nosferatu	
Dracula nycterina	
Dracula octavioi	
Dracula ophioceps	
Dracula orientalis	
Dracula ortiziana	
Dracula papillosa	
Dracula pholeodytes	
Dracula × pileus	
Dracula platycrater	
Dracula polyphemus	
Dracula portillae	
Dracula posadarum	
Dracula presbys	
Dracula psittacina	
Dracula psyche	
Dracula pubescens	
Dracula pusilla	
Dracula quilichaoensis	Dracula trichroma
Dracula radiella	
Dracula radiosa	
Dracula × radio-syndactyla	
Dracula rezekiana	
Dracula ripleyana	
Dracula robledorum	
Dracula roezlii	
Dracula senilis	Dracula chimaera
Dracula sergioi	
Dracula severa	
Dracula sibundoyensis	
Dracula simia	
Dracula sodiroi	
Dracula syndactyla	
Dracula tarantula	Dracula tubeana
Dracula trichroma	

*For explanation see page 4, point 6
*Voir les explications page 9, point 6
*Para mayor explicación, véase la página 15, point 6

ALL NAMES	ACCEPTED NAMES
Dracula trinema	Dracula platycrater
Dracula trinema	Dracula velutina
Dracula trinympharum	
Dracula troglodytes	Dracula benedictii
Dracula tubeana	
Dracula ubangina	
Dracula vagabunda	Dracula pusilla
Dracula vampira	
Dracula velutina	
Dracula venefica	
Dracula venosa	
Dracula verticulosa	
Dracula vespertilio	
Dracula vinacea	
Dracula vlad-tepes	
Dracula wallisii	
Dracula woolwardiae	
Dracula xenos	
Encyclia abbreviata	
Encyclia acicularis	Encyclia bractescens
Encyclia acuta	
Encyclia adenocarpon	
Encyclia adenocarpon subsp. *trachycarpa*	Encyclia trachycarpa
Encyclia adenocaula	
Encyclia adenocaula var. **kennedyi**	
Encyclia advena	
Encyclia aemula	
Encyclia aenicta	
Encyclia alagoensis	
Encyclia alanjensis	Encyclia stellata
Encyclia alata	
Encyclia alata subsp. *parvifera*	Encyclia parviflora
Encyclia alata var. *parviflora*	Encyclia parviflora
Encyclia alata var. *virella*	Encyclia alata
Encyclia albopurpurea	
Encyclia alboxanthina	
Encyclia allemanii	
Encyclia allemanoides	
Encyclia amabilis	Encyclia concolor
Encyclia amanda	
Encyclia ambigua	
Encyclia amicta	
Encyclia andrichii	
Encyclia angustiloba	
Encyclia apuahuensis	
Encyclia argentinensis	
Encyclia arminii	
Encyclia aromatica	Encyclia incumbens
Encyclia aspera	
Encyclia asperirachis	
Encyclia asperula	
Encyclia atropurpurea	Encyclia cordigera
Encyclia atropurpurea var. *rosea*	Encyclia cordigera
Encyclia atropurpureum var. *roseum*	Encyclia hanburii
Encyclia atropururea var. *leucantha*	Encyclia cordigera
Encyclia atropururea var. *rhodoglossa*	Encyclia cordigera

*For explanation see page 4, point 6
*Voir les explications page 9, point 6
*Para mayor explicación, véase la página 15, point 6

Part I: All Names / Tous les Noms / Todos los Nombres

ALL NAMES	ACCEPTED NAMES
Encyclia atrorubens	
Encyclia auyantepuiensis	
Encyclia baculus	
Encyclia bahamensis	Encyclia rufa
Encyclia bahiensis	Encyclia fowliei
Encyclia belizensis	Encyclia guatemalensis
Encyclia bicamerata	
Encyclia bicornuta	Encyclia linearifolioides
Encyclia bipapularis	Encyclia triangulifera
Encyclia boothiana	
Encyclia boothiana subsp. boothiana	
Encyclia boothiana subsp. favoris	
Encyclia boothiana var. *erythronioides*	Encyclia boothiana
Encyclia brachiata	
Encyclia brachychila	Encyclia hartwegii
Encyclia bracteata	
Encyclia bractescens	
Encyclia bradfordi	
Encyclia bragancae	
Encyclia brassavolae	
Encyclia brenesii	Encyclia mooreana
Encyclia buchtienii	
Encyclia bulbosa	Encyclia inversa
Encyclia burle-marxii	
Encyclia caetensis	
Encyclia calamaria	
Encyclia campos-portoi	
Encyclia campylostalix	
Encyclia candollei	
Encyclia cardimii	
Encyclia carpatiana	
Encyclia cepiforme	
Encyclia ceratistes	
Encyclia chacaoensis	
Encyclia chapadensis	
Encyclia chiapasensis	
Encyclia chimborazoensis	
Encyclia chiriquensis	Encyclia varicosa
Encyclia chiriquensis	Encyclia varicosa subsp. varicosa
Encyclia chloroleuca	
Encyclia chondylobulbon	
Encyclia citrina	
Encyclia cochleata	
Encyclia conchaechila	
Encyclia concolor	
Encyclia confusa	
Encyclia conspicua	Encyclia dichroma
Encyclia cordigera	
Encyclia cordigera var. *rosea*	Encyclia cordigera
Encyclia cretacea	
Encyclia cyanocolumna	
Encyclia cyperifolia	
Encyclia deamii	Encyclia livida
Encyclia dichroma	
Encyclia dickinsoniana	
Encyclia diguetii	Encyclia tripunctata

*For explanation see page 4, point 6
*Voir les explications page 9, point 6
*Para mayor explicación, véase la página 15, point 6

ALL NAMES	ACCEPTED NAMES
Encyclia diota	
Encyclia diota subsp. *atrorubens*	Encyclia atrorubens
Encyclia diota subsp. *diota*	Encyclia atrorubens
Encyclia distantiflora	
Encyclia diurna	
Encyclia doeringii	
Encyclia dutrai	
Encyclia duveenii	
Encyclia ensiformis	
Encyclia ensiformis	Encyclia vellozoana
Encyclia euosma	
Encyclia faresiana	
Encyclia fausta	
Encyclia favoris	Encyclia boothiana subsp. favoris
Encyclia flabellata	Encyclia candollei
Encyclia flabellifera	
Encyclia flava	
Encyclia fortunae	
Encyclia fowliei	
Encyclia fragrans	Encyclia aemula
Encyclia fragrans	
Encyclia fragrans subsp. *aemula*	Encyclia fragrans
Encyclia fragrans var. *brevistriatum*	Encyclia fragrans
Encyclia gallopavina	
Encyclia garciana	
Encyclia ghiesbreghtiana	
Encyclia ghillanyi	
Encyclia glauca	
Encyclia gonzalezii	
Encyclia goyazensis	
Encyclia grammatoglossa	
Encyclia granitica	
Encyclia gravida	
Encyclia guatemalensis	
Encyclia guianensis	
Encyclia guttata	Encyclia maculosa
Encyclia hanburii	
Encyclia hartwegii	
Encyclia hastata	
Encyclia hoehnei	
Encyclia hoffmanii	Encyclia chacoensis
Encyclia hollandiae	
Encyclia huebneri	
Encyclia hunteriana	Encyclia stellata
Encyclia icthyphylla	Encyclia michuacana
Encyclia incumbens	
Encyclia insidiosa	Encyclia diota
Encyclia inversa	
Encyclia ionophlebia	Encyclia chacaoensis
Encyclia ionophlebium	
Encyclia ionosma	
Encyclia ivonae	
Encyclia jauana	
Encyclia jenischiana	Encyclia dichroma
Encyclia kautskyi	
Encyclia kennedyi	

*For explanation see page 4, point 6
*Voir les explications page 9, point 6
*Para mayor explicación, véase la página 15, point 6

ALL NAMES	ACCEPTED NAMES
Encyclia kienastii	
Encyclia lambda	
Encyclia lancifolia	
Encyclia latipetala	
Encyclia laxa	Encyclia candollei
Encyclia leucantha	
Encyclia limbata	Encyclia glauca
Encyclia lindenii	
Encyclia linearifolioides	
Encyclia linearis	Encyclia luteorosea
Encyclia linkiana	
Encyclia livida	
Encyclia longifolia	
Encyclia lorata	
Encyclia luteorosea	
Encyclia lutzenbergerii	
Encyclia macrochila	Encyclia cordigera
Encyclia maculosa	
Encyclia maderoi	
Encyclia magnispatha	
Encyclia mapuerae	
Encyclia mariae	
Encyclia megalantha	
Encyclia megalantha var. *spiritusantensis*	Encyclia spiritusanctensis
Encyclia meliosma	
Encyclia michuacana	
Encyclia microbulbon	
Encyclia microtos	
Encyclia microxanthina	
Encyclia moebusii	Encyclia triangulifera
Encyclia moojenii	
Encyclia mooreana	
Encyclia multiflora	Encyclia viridiflora
Encyclia naranjapatensis	
Encyclia nematocaulon	
Encyclia nemoralis	Encyclia adenocaula
Encyclia neurosa	
Encyclia obpyribulbon	
Encyclia ochracea	
Encyclia oestlundii	
Encyclia oncidioides	
Encyclia oncidioides var. *gravida*	Encyclia gravida
Encyclia oncidioides var. *ramonensis*	Encyclia ceratistes
Encyclia organensis	Encyclia calamaria
Encyclia osmantha	
Encyclia ovulum	Encyclia microbulbon
Encyclia oxiphylla	
Encyclia pachyantha	
Encyclia pamplonense	Encyclia tigrina
Encyclia panthera	
Encyclia papilio	
Encyclia papillosa	
Encyclia parviflora	
Encyclia pastons	Encyclia venosa
Encyclia pastoris	Encyclia linkiana
Encyclia patens	

*For explanation see page 4, point 6
*Voir les explications page 9, point 6
*Para mayor explicación, véase la página 15, point 6

ALL NAMES	ACCEPTED NAMES
Encyclia pauciflora	
Encyclia pedra-azulensis	
Encyclia pentotis	Encyclia baculus
Encyclia peraltense	
Encyclia peraltensis	Encyclia ceratistes
Encyclia perplexa	
Encyclia × perplexa	Encyclia perplexa
Encyclia pflanzii	
Encyclia picta	
Encyclia pipio	
Encyclia playmatoglossa	Encyclia varicosa
Encyclia playmatoglossa	Encyclia varicosa subsp. varicosa
Encyclia pollardiana	
Encyclia polybulbon	
Encyclia porrecta	
Encyclia powellii	Encyclia ceratistes
Encyclia pringlei	
Encyclia prismatocarpa	
Encyclia pruinosa	Encyclia concolor
Encyclia pseudopygmaea	
Encyclia pterocarpa	
Encyclia pulcherrima	
Encyclia punctifera	
Encyclia purpusii	Encyclia nematocaulon
Encyclia pygmaea	
Encyclia radiata	
Encyclia ramonense	Encyclia ceratistes
Encyclia randii	
Encyclia recurvata	
Encyclia regnellania	
Encyclia rhombilabia	
Encyclia rhynchophora	
Encyclia rufa	
Encyclia saltensis	Encyclia argentinensis
Encyclia sceptra	
Encyclia schmidtii	
Encyclia sclerocladia	
Encyclia seidelii	
Encyclia selligera	
Encyclia semiaperta	
Encyclia serroniana	Encyclia patens
Encyclia sessiflora	
Encyclia sima	
Encyclia sisyrinchiifolia	Encyclia microbulbon
Encyclia spatella	
Encyclia spiritusanctensis	
Encyclia spondiadum	
Encyclia squamata	Encyclia hoehnei
Encyclia steinbachii	
Encyclia stellata	
Encyclia suaveolens	
Encyclia subulatifolia	
Encyclia suzanensis	
Encyclia tampensis	
Encyclia tarumana	
Encyclia tenuissima	

*For explanation see page 4, point 6
*Voir les explications page 9, point 6
*Para mayor explicación, véase la página 15, point 6

Part I: All Names / Tous les Noms / Todos los Nombres

ALL NAMES	ACCEPTED NAMES
Encyclia tesselata	Encyclia livida
Encyclia thienii	Encyclia chloroleuca
Encyclia tigrina	
Encyclia tonduziana	Encyclia mooreana
Encyclia trachycarpa	
Encyclia trachychila	Encyclia ambigua
Encyclia triangulifera	
Encyclia tripartita	
Encyclia triptera	Encyclia pygmaea
Encyclia tripterum	Encyclia semiaperta
Encyclia tripunctata	
Encyclia tuerckheimii	
Encyclia unaensis	
Encyclia vagans	
Encyclia varicosa	
Encyclia varicosa subsp. leiobulbon	
Encyclia varicosa subsp. varicosa	
Encyclia vellozoana	
Encyclia venezuelana	
Encyclia venosa	
Encyclia vespa	
Encyclia virens	
Encyclia virgata	Encyclia michuacana
Encyclia viridiflava	
Encyclia viridiflora	
Encyclia vitellina	
Encyclia wageneri	Encyclia diurna
Encyclia wendlandiana	Encyclia venosa
Encyclia widgrenii	
Encyclia xerophytica	
Encyclia xipheres	Encyclia nematocaulon
Encyclia xipheroides	
Encyclia xuxiana	
Encyclia yauperyensis	
Encylia glumacea	
Epicladium boothianum	Encyclia boothiana
Epicladium boothianum var. *erythronioides*	Encyclia boothiana
Epidendrum abbreviatum	Encyclia abbreviata
Epidendrum aciculare	Encyclia bractescens
Epidendrum acuminatum	Encyclia baculus
Epidendrum acutum	Encyclia acuta
Epidendrum adenocarpon	Encyclia adenocarpon
Epidendrum adenocarpon subsp. *papillosa*	Encyclia adenocarpon
Epidendrum adenocarpon var. *rosei*	Encyclia adenocarpon
Epidendrum adenocaulum	Encyclia adenocaula
Epidendrum advena	Encyclia advena
Epidendrum aemulum	Encyclia aemula
Epidendrum alagoense	Encyclia alagoense
Epidendrum alanjense	Encyclia stellata
Epidendrum alatum	Encyclia alata
Epidendrum alatum var. *arrogans*	Encyclia alata
Epidendrum alatum var. *grandiflorum*	Encyclia alata
Epidendrum alatum var. *longipetalum*	Encyclia alata
Epidendrum alatum var. *viridiflorum*	Encyclia alata
Epidendrum albopurpureum	Encyclia albopurpurea
Epidendrum allemanii	Encyclia allemanii

*For explanation see page 4, point 6
*Voir les explications page 9, point 6
*Para mayor explicación, véase la página 15, point 6

ALL NAMES	ACCEPTED NAMES
Epidendrum allemanoides	Encyclia allemanoides
Epidendrum almasyi	Encyclia glumacea
Epidendrum aloides	Cymbidium aloifolium
Epidendrum aloifolium	Cymbidium aloifolium
Epidendrum amabile	Encyclia concolor
Epidendrum amabile	Encyclia dichroma
Epidendrum ambiguum	Encyclia ambigua
Epidendrum amictum	Encyclia amicta
Epidendrum apuahuense	Encyclia apuahuensis
Epidendrum argentinense	Encyclia argentinensis
Epidendrum arminii	Encyclia arminii
Epidendrum aromaticum	Encyclia incumbens
Epidendrum articulatum	Encyclia livida
Epidendrum asperum	Encyclia aspera
Epidendrum asperum	Encyclia candollei
Epidendrum atropurpureum	Encyclia cordigera
Epidendrum atropurpureum var. *roseum*	Encyclia cordigera
Epidendrum atropurureum var. *laciniatum*	Encyclia cordigera
Epidendrum atropurureum var. *lionetianum*	Encyclia cordigera
Epidendrum atropurureum var. *longilabre*	Encyclia cordigera
Epidendrum atrorubens	Encyclia atrorubens
Epidendrum baculibulbum	Encyclia vespa
Epidendrum baculus	Encyclia baculus
Epidendrum bahamense	Encyclia rufa
Epidendrum belizense	Encyclia guatemalensis
Epidendrum beyrodtianum	Encyclia baculus
Epidendrum bicameratum	Encyclia bicamerata
Epidendrum bidentatum	Encyclia boothiana
Epidendrum biflorum	Encyclia dichroma
Epidendrum bipapulare	Encyclia triangulifera
Epidendrum boothianum	Encyclia boothiana
Epidendrum brachiatum	Encyclia brachiata
Epidendrum brachychilum	Encyclia hartwegii
Epidendrum bracteatum	Encyclia bracteata
Epidendrum bractescens	Encyclia bractescens
Epidendrum bradfordii	Encyclia bradfordi
Epidendrum brassavolae	Encyclia brassavolae
Epidendrum bulbosum	Encyclia inversa
Epidendrum caespitosum	Encyclia pygmaea
Epidendrum calamarium	Encyclia calamaria
Epidendrum callista	Dendrobium amabile
Epidendrum calocheilum	Encyclia alata
Epidendrum campylostalix	Encyclia campylostalix
Epidendrum candollei	Encyclia candollei
Epidendrum carpatianum	Encyclia carpatiana
Epidendrum ceratistes	Encyclia ceratistes
Epidendrum chacaoense	Encyclia chacaoensis
Epidendrum chimborazoense	Encyclia chimborazoensis
Epidendrum chiriquense	Encyclia varicosa
Epidendrum chiriquense	Encyclia varicosa subsp. varicosa
Epidendrum chloroleucum	Encyclia chloroleuca
Epidendrum chondylobulbon	Encyclia chondylobulbon
Epidendrum christi	Encyclia vespa
Epidendrum cinnamomeum	Encyclia pterocarpa
Epidendrum citrinum	Encyclia citrina
Epidendrum cochleatum	Encyclia cochleata

*For explanation see page 4, point 6
*Voir les explications page 9, point 6
*Para mayor explicación, véase la página 15, point 6

ALL NAMES	ACCEPTED NAMES
Epidendrum cochleatum var. *arrogans*	Encyclia cochleata
Epidendrum cochleatum var. *costaricense*	Encyclia cochleata
Epidendrum cochleatum var. *pallidum*	Encyclia cochleata
Epidendrum cochleatum var. *trandrum*	Encyclia cochleata
Epidendrum conchaechilum	Encyclia conchaechila
Epidendrum concolor	Encyclia concolor
Epidendrum condylochilum	Encyclia livida
Epidendrum confusum	Encyclia baculus
Epidendrum conspicuum	Encyclia dichroma
Epidendrum cordigerum	Encyclia cordigera
Epidendrum coriaceum	Encyclia vespa
Epidendrum crassilabium	Encyclia vespa
Epidendrum crispatum	Encyclia adenocarpon
Epidendrum cyanocolumna	Encyclia cyanocolumna
Epidendrum cyperifolium	Encyclia cyperifolia
Epidendrum dasytaenia	Encyclia livida
Epidendrum deamii	Encyclia livida
Epidendrum dichromum	Encyclia dichroma
Epidendrum dickinsonianum	Encyclia dickinsoniana
Epidendrum diguetii	Encyclia tripunctata
Epidendrum diotum	Encyclia diota
Epidendrum distantiflorum	Encyclia distantiflora
Epidendrum diurnum	Encyclia diurna
Epidendrum ensicaulon	Encyclia venosa
Epidendrum ensifolium	Cymbidium ensifolium
Epidendrum ensiforme	Encyclia vellozoana
Epidendrum erythronioides	Encyclia boothiana
Epidendrum esculentum	Encyclia bractescens
Epidendrum euosmum	Encyclia euosma
Epidendrum falax	Encyclia lindenii
Epidendrum fallax var. *flavecens*	Encyclia lindenii
Epidendrum faustum	Encyclia fausta
Epidendrum feddeanum	Encyclia vespa
Epidendrum flabellatum	Encyclia candollei
Epidendrum flava	Encyclia flava
Epidendrum formosum	Encyclia alata
Epidendrum fragrans	Encyclia aemula
Epidendrum fragrans	Encyclia calamaria
Epidendrum fragrans	Encyclia fragrans
Epidendrum fragrans	Encyclia inversa
Epidendrum fragrans var. *megalanthum*	Encyclia baculus
Epidendrum fragrans var. *pachypus*	Encyclia fragrans
Epidendrum fuscum	Encyclia hartwegii
Epidendrum gallopavinum	Encyclia gallopavina
Epidendrum garcianum	Encyclia garciana
Epidendrum ghiesbreghtianum	Encyclia ghiesbreghtiana
Epidendrum glaucovirens	Encyclia glauca
Epidendrum glaucum	Encyclia glauca
Epidendrum glumaceum	Encyclia glumacea
Epidendrum glutinosum	Encyclia patens
Epidendrum godseffianum	Encyclia carpatiana
Epidendrum grammatoglossum	Encyclia grammatoglossa
Epidendrum graniticum	Encyclia granitica
Epidendrum gravidum	Encyclia gravida
Epidendrum guatemalense	Encyclia guatemalensis
Epidendrum guttatum	Encyclia maculosa

*For explanation see page 4, point 6
*Voir les explications page 9, point 6
*Para mayor explicación, véase la página 15, point 6

ALL NAMES	ACCEPTED NAMES
Epidendrum hallatum	Encyclia recurvata
Epidendrum hanburii	Encyclia hanburii
Epidendrum hartwegii	Encyclia hartwegii
Epidendrum hastatum	Encyclia hastata
Epidendrum henrici	Encyclia livida
Epidendrum hoehnei	Encyclia hoehnei
Epidendrum icthyphyllum	Encyclia michuacana
Epidendrum incumbens	Encyclia incumbens
Epidendrum insidiosum	Encyclia diota
Epidendrum inversum	Encyclia inversa
Epidendrum ionocentrum	Encyclia prismatocarpa
Epidendrum ionophlebium	Encyclia ionophlebium
Epidendrum ionosmum	Encyclia ionosma
Epidendrum jenischianum	Encyclia dichroma
Epidendrum karwiushii	Encyclia bicamerata
Epidendrum kennedyi	Encyclia kennedyi
Epidendrum kienastii	Encyclia kienastii
Epidendrum lambda	Encyclia lambda
Epidendrum lancifolium	Encyclia lancifolia
Epidendrum langlassei	Encyclia lancifolia
Epidendrum latipetalum	Encyclia latipetala
Epidendrum latro	Encyclia inversa
Epidendrum leiobulbon	Encyclia varicosa var. leiobulbon
Epidendrum leopardinum	Encyclia vespa
Epidendrum leucanthum	Encyclia leucantha
Epidendrum limbatum	Encyclia glauca
Epidendrum lindenii	Encyclia lindenii
Epidendrum lineare	Encyclia luteorosea
Epidendrum linearifolioides	Encyclia linearifolioides
Epidendrum linearifolium	Encyclia bractescens
Epidendrum lineatum	Encyclia fragrans
Epidendrum linkianum	Encyclia linkiana
Epidendrum lividum	Encyclia livida
Epidendrum longifolium	Encyclia longifolia
Epidendrum longipes	Encyclia vespa
Epidendrum longipetalum	Encyclia alata
Epidendrum longipetalum	Encyclia cordigera*
Epidendrum luteo-roseum	Encyclia luteorosea
Epidendrum macrochilum	Encyclia cordigera
Epidendrum macrochilum var. *albopurpureum*	Encyclia cordigera
Epidendrum macrochilum var. *roseum*	Encyclia cordigera
Epidendrum maculosum	Encyclia maculosa
Epidendrum madrense	Encyclia chacaoensis
Epidendrum magnispathum	Encyclia magnispatha
Epidendrum marginatum	Encyclia radiata
Epidendrum mariae	Encyclia mariae
Epidendrum megalanthum	Encyclia megalantha
Epidendrum meliosmum	Encyclia meliosma
Epidendrum michuacanum	Encyclia michuacana
Epidendrum microbulbon	Encyclia microbulbon
Epidendrum micropus	Encyclia tripunctata
Epidendrum microtos	Encyclia microtos
Epidendrum microtos var. *grandiflorum*	Encyclia cyperifolia
Epidendrum monanthum	Encyclia pygmaea
Epidendrum monile	Dendrobium moniliforme
Epidendrum moniliforme	Dendrobium moniliforme

*For explanation see page 4, point 6
*Voir les explications page 9, point 6
*Para mayor explicación, véase la página 15, point 6

ALL NAMES	ACCEPTED NAMES
Epidendrum moojenii	**Encyclia moojenii**
Epidendrum mooreanum	**Encyclia mooreana**
Epidendrum moschatum	**Dendrobium moschatum**
Epidendrum nematocaulon	**Encyclia nematocaulon**
Epidendrum nemorale	**Encyclia adenocaula**
Epidendrum neurosum	**Encyclia neurosa**
Epidendrum ochraceum	**Encyclia ochracea**
Epidendrum ochranthum	**Encyclia diurna**
Epidendrum odoratissima	**Encyclia patens**
Epidendrum odoratissimum	**Encyclia patens**
Epidendrum oestlundii	**Encyclia oestlundii**
Epidendrum oncidioides	**Encyclia oncidioides**
Epidendrum oncidioides var. *gravidum*	**Encyclia gravida**
Epidendrum oncidioides var. *mooreanum*	**Encyclia mooreana**
Epidendrum oncidioides var. *perplexum*	**Encyclia perplexa**
Epidendrum oncidioides var. *ramonense*	**Encyclia ceratistes**
Epidendrum oncioides var. *gravida*	**Encyclia gravida**
Epidendrum organense	**Encyclia calamaria**
Epidendrum osmanthum	**Encyclia osmantha**
Epidendrum ovulum	**Encyclia microbulbon**
Epidendrum oxiphyllum	**Encyclia oxiphylla**
Epidendrum pachyanthum	**Encyclia hartwegii**
Epidendrum pachyanthum	**Encyclia pachyantha**
Epidendrum pachycarpum	**Encyclia chacaoensis**
Epidendrum pachysepalum	**Encyclia vespa**
Epidendrum pamplonense	**Encyclia tigrina**
Epidendrum panthera	**Encyclia panthera**
Epidendrum papilio	**Encyclia papilio**
Epidendrum papillosum	**Encyclia adenocarpon**
Epidendrum papyriferum	**Encyclia panthera**
Epidendrum parviflorum	**Encyclia ochracea**
Epidendrum pastons	**Encyclia venosa**
Epidendrum pastoris	**Encyclia linkiana**
Epidendrum pauciflorum	**Encyclia pauciflora**
Epidendrum pendulum	**Cymbidium aloifolium**
Epidendrum pentotis	**Encyclia baculus**
Epidendrum peraltense	**Encyclia peraltense**
Epidendrum pflanzii	**Encyclia pflanzii**
Epidendrum phymatoglossum	**Encyclia varicosa**
Epidendrum phymatoglossum	**Encyclia varicosa** subsp. varicosa
Epidendrum pictum	**Encyclia picta**
Epidendrum pipio	**Encyclia pipio**
Epidendrum platycardium	**Encyclia spondiadum**
Epidendrum pollardianum	**Encyclia aenicta**
Epidendrum polybulbon	**Encyclia polybulbon**
Epidendrum polybulbon var. *luteo-album*	**Encyclia polybulbon**
Epidendrum primullinum	**Encyclia rufa**
Epidendrum primuloides	**Encyclia incumbens**
Epidendrum pringlei	**Encyclia pringlei**
Epidendrum prismatocarpum	**Encyclia prismatocarpa**
Epidendrum prorepens	**Encyclia abbreviata**
Epidendrum pruinosum	**Encyclia concolor**
Epidendrum pterocarpum	**Encyclia pterocarpa**
Epidendrum punctiferum	**Encyclia punctifera**
Epidendrum punctulatum	**Encyclia concolor**
Epidendrum purpurachylum	**Encyclia gallopavina**

*For explanation see page 4, point 6
*Voir les explications page 9, point 6
*Para mayor explicación, véase la página 15, point 6

72

ALL NAMES

Epidendrum pusillum
Epidendrum pygmaeum
Epidendrum quadidentatum
Epidendrum quadratum
Epidendrum quadratum
Epidendrum radiatum
Epidendrum ramírezzi
Epidendrum ramonense
Epidendrum randii
Epidendrum regnellianum
Epidendrum rhabdobulbon
Epidendrum rhopalobulbon
Epidendrum rhynchophorum
Epidendrum rueckerae
Epidendrum rufum
Epidendrum saccharatum
Epidendrum sceptrum
Epidendrum sclerocladium
Epidendrum selligerum
Epidendrum seriatum
Epidendrum serronianum
Epidendrum sessiflorum
Epidendrum simum
Epidendrum sinense
Epidendrum sinense
Epidendrum sisyrinchiifolium
Epidendrum spatella
Epidendrum spatulatum
Epidendrum spondiadum
Epidendrum squamatum
Epidendrum stellatum
Epidendrum subulatifolium
Epidendrum suzanense
Epidendrum tampense
Epidendrum tampense var. *amesianum*
Epidendrum tarumanum
Epidendrum tenuissimum
Epidendrum tessellatum
Epidendrum tigrinum
Epidendrum trachycarpum
Epidendrum trachychilum
Epidendrum triandrum
Epidendrum triangulifera
Epidendrum tripartitum
Epidendrum tripterum
Epidendrum tripterum
Epidendrum tripunctatum
Epidendrum tripunctatum
Epidendrum triste
Epidendrum trulla
Epidendrum tuerckheimii
Epidendrum uniflorum
Epidendrum vagans
Epidendrum vaginatum
Epidendrum varicosum
Epidendrum varicosum

ACCEPTED NAMES

Encyclia bracteata
Encyclia pygmaea
Encyclia grammatoglossa
Encyclia varicosa
Encyclia varicosa subsp. **varicosa**
Encyclia radiata
Encyclia varicosa subsp. **leiobulbon**
Encyclia ceratistes
Encyclia randii
Encyclia regnelliana
Encyclia vespa
Encyclia vespa
Encyclia rhynchophora
Encyclia lambda
Encyclia rufa
Encyclia vespa
Encyclia sceptra
Encyclia sclerocladia
Encyclia selligera
Encyclia luteorosea
Encyclia patens
Encyclia sessiflora
Encyclia sima
Cymbidium ensifolium
Cymbidium sinense
Encyclia microbulbon
Encyclia spatella
Dendrobium crumenatum
Encyclia spondiadum
Encyclia hoehnei
Encyclia stellata
Encyclia subulatifolia
Encyclia suzanensis
Encyclia tampensis
Encyclia triangulifera
Encyclia tarumana
Encyclia tenuissima
Encyclia livida
Encyclia tigrina
Encyclia trachycarpa
Encyclia ambigua
Encyclia cochleata
Encyclia triangulifera
Encyclia tripartita
Encyclia linkiana
Encyclia pygmaea
Encyclia punctifera
Encyclia tripunctata
Encyclia ochracea
Encyclia lancifolia
Encyclia tuerckheimii
Encyclia pygmaea
Encyclia vagans
Encyclia fragrans
Encyclia varicosa
Encyclia varicosa subsp. **varicosa**

*For explanation see page 4, point 6
*Voir les explications page 9, point 6
*Para mayor explicación, véase la página 15, point 6

Part I: All Names / Tous les Noms / Todos los Nombres

ALL NAMES	ACCEPTED NAMES
Epidendrum variegatum	Encyclia vespa
Epidendrum venezuelanum	Encyclia venezuelana
Epidendrum venosum	Encyclia venosa
Epidendrum verrucosum	Encyclia adenocaula
Epidendrum vespa	Encyclia vespa
Epidendrum violodora	Encyclia selligera
Epidendrum virens	Encyclia virens
Epidendrum virgatum	Encyclia michuacana
Epidendrum virgatum var. *arrogans*	Encyclia michuacana
Epidendrum viridiflorum	Encyclia viridiflora
Epidendrum vitellinum	Encyclia vitellina
Epidendrum wageneri	Encyclia diurna
Epidendrum wendlandianum	Encyclia venosa
Epidendrum widgrenii	Encyclia widgrenii
Epidendrum xipheres	Encyclia nematocaulon
Epigeneium simplex	Dendrobium simplex
Epithecia glauca	Encyclia glauca
Grammatophyllum elegans	Cymbidium elegans
Hormidium allemanii	Encyclia allemanii
Hormidium allemanoides	Encyclia allemanoides
Hormidium almasyi	Encyclia glumacea
Hormidium boothianum	Encyclia boothiana
Hormidium caetense	Encyclia caetensis
Hormidium calamarium	Encycli calamaria
Hormidium campos-portoi	Encyclia campos-portoi
Hormidium cochleatum	Encyclia cochleata
Hormidium faresianum	Encyclia faresiana
Hormidium glumaceum	Encyclia glumacea
Hormidium gramatoglossum	Encyclia grammatoglossa
Hormidium hartwegii	Encyclia hartwegii
Hormidium humile	Encyclia pygmaea
Hormidium inversum	Encyclia inversa
Hormidium lividum	Encyclia livida
Hormidium papilio	Encyclia papilio
Hormidium prismatocarpum	Encyclia prismatocarpa
Hormidium pseudopygmaeum	Encyclia pseudopygmaea
Hormidium pygmaeum	Encyclia pygmaea
Hormidium tripterum	Encyclia pygmaea
Hormidium uniflorum	Encyclia pygmaea
Hormidium widgrenii	Encyclia widgrenii
Jensoa ensata	Cymbidium ensifolium
Katherinea carrii	Dendrobium carrii
Katherinea parvula	Dendrobium delicatulum var. parvulum
Katherinea simplex	Dendrobium simplex
Katherinea uncipes	Dendrobium uncipes
Latourea spectabilis	Dendrobium spectabile
Latourorchis alexandrae	Dendrobium alexandrae
Latourorchis atroviolaceum	Dendrobium atroviolaceum
Latourorchis forbesii	Dendrobium forbesii
Latourorchis leucohybos	Dendrobium leucohybos
Latourorchis macrophylla	Dendrobium macrophyllum
Latourorchis muscifera	Dendrobium macrophyllum
Latourorchis spectabile	Dendrobium spectabile
Limodorum aphyllum	Dendrobium aphyllum
Limodorum diurnum	Encyclia diurna

*For explanation see page 4, point 6
*Voir les explications page 9, point 6
*Para mayor explicación, véase la página 15, point 6

74

ALL NAMES	ACCEPTED NAMES
Limodorum ensatum	**Cymbidium ensifolium**
Limodorum longicorne	**Disa filicornis**
Macrostomium aloefolium	**Dendrobium aloifolium**
Masdevallia andreettae	**Dracula andreettae**
Masdevallia astuta	**Dracula erythrochaete** subsp. **astuta**
Masdevallia astuta var. *gaskelliana*	**Dracula erythrochaete**
Masdevallia backhousiana	**Dracula chimaera**
Masdevallia bella	**Dracula bella**
Masdevallia benedictii	**Dracula benedictii**
Masdevallia bomboiza	**Dracula lotax**
Masdevallia burbidgeana	**Dracula erythrochaete**
Masdevallia callifera	**Dracula houtteana**
Masdevallia carderi	**Dracula inaequalis**
Masdevallia carderi var. *mosquerae*	**Dracula houtteana**
Masdevallia carderiopsis	**Dracula houtteana**
Masdevallia chestertonii	**Dracula chestertonii**
Masdevallia chimaera	**Dracula chimaera***
Masdevallia chimaera	**Dracula nycterina**
Masdevallia chimaera var. *gorgona*	**Dracula gorgona**
Masdevallia chimaera var. *robledorum*	**Dracula robledorum**
Masdevallia chimaera var. *backhousiana*	**Dracula chimaera**
Masdevallia chimaera var. *roezlii*	**Dracula roezlii**
Masdevallia chimaera var. *senilis*	**Dracula chimaera**
Masdevallia chimaera var. *severa*	**Dracula severa**
Masdevallia chimaera var. *wallisii*	**Dracula wallisii**
Masdevallia chimaera var. *winniana*	**Dracula roezlii**
Masdevallia chimaera var. *aurantiaca*	**Dracula chimaera**
Masdevallia deltoidea	**Dracula deltoidea**
Masdevallia dodsonii	**Dracula dodsonii**
Masdevallia erythrochaete	**Dracula erythrochaete**
Masdevallia erythrochaete var. *astuta*	**Dracula erythrochaete** subsp. **astuta**
Masdevallia erythrochaete var. *gaskelliana*	**Dracula erythrochaete**
Masdevallia felix	**Dracula felix**
Masdevallia fuliginosa	**Dracula radiella**
Masdevallia gaskelliana	**Dracula erythrochaete**
Masdevallia gigas	**Dracula gigas**
Masdevallia gorgo	**Dracula erythrochaete** subsp. **astuta**
Masdevallia houtteana	**Dracula houtteana**
Masdevallia inaequalis	**Dracula inaequalis**
Masdevallia iricolor	**Dracula trichroma**
Masdevallia janetiae	**Dracula janetiae**
Masdevallia johannis	**Dracula pusilla**
Masdevallia lactea	**Dracula velutina**
Masdevallia lotax	**Dracula lotax**
Masdevallia lowii	**Dracula platycrater**
Masdevallia macrochila	**Dracula chestertonii**
Masdevallia medellinensis	**Dracula radiosa**
Masdevallia microglochin	**Dracula velutina**
Masdevallia mopsus	**Dracula mopsus**
Masdevallia mosquerae	**Dracula houtteana**
Masdevallia nycterina	**Dracula nycterina**
Masdevallia platycrater	**Dracula platycrater**
Masdevallia polyphemus	**Dracula polyphemus**
Masdevallia psittacina	**Dracula psittacina**
Masdevallia psyche	**Dracula psyche**
Masdevallia pusilla	**Dracula pusilla**

*For explanation see page 4, point 6
*Voir les explications page 9, point 6
*Para mayor explicación, véase la página 15, point 6

Part I: All Names / Tous les Noms / Todos los Nombres

ALL NAMES	ACCEPTED NAMES
Masdevallia quilichaoensis	**Dracula trichroma**
Masdevallia radiosa	**Dracula radiosa**
Masdevallia roezlii	**Dracula roezlii**
Masdevallia roezlii var. *rubra*	**Dracula roezlii**
Masdevallia roezlii var. *rubrum*	**Dracula roezlii**
Masdevallia senilis	**Dracula chimaera**
Masdevallia severa	**Dracula severa**
Masdevallia sodiroi	**Dracula sodiroi**
Masdevallia spectrum	**Dracula severa**
Masdevallia tarantula	**Dracula tubeana**
Masdevallia triceratops	**Dracula mopsus**
Masdevallia trichroma	**Dracula trichroma**
Masdevallia tricolor	**Dracula trichroma**
Masdevallia trinema	**Dracula platycrater***
Masdevallia trinema	**Dracula velutina**
Masdevallia troglodytes	**Dracula benedictii**
Masdevallia tubeana	**Dracula tubeana**
Masdevallia vampira	**Dracula vampira**
Masdevallia velutina	**Dracula velutina**
Masdevallia venosa	**Dracula venosa**
Masdevallia vespertilio	**Dracula vespertilio**
Masdevallia wallisii	**Dracula wallisii**
Masdevallia wallisii var. *discoidea*	**Dracula wallisii**
Masdevallia wallisii var. *stupenda*	**Dracula chimaera**
Masdevallia winniana	**Dracula roezlii**
Masdevallia woolwardiae	**Dracula woolwardiae**
Maxillaria goeringii	**Cymbidium goeringii**
Microstylis humilis	**Encyclia pygmaea**
Monadenia basutorum	**Disa basutorum**
Monadenia junodiana	**Disa fragrans** subsp. **fragrans**
Monadenia leydenburgensis	**Disa stachyoides**
Monadenia tenuis	**Disa tenuis**
Onychium tetraedre	**Dendrobium tetraedre**
Onychium affine	**Dendrobium affine**
Onychium crumenatum	**Dendrobium crumenatum**
Onychium fimbriatum	**Dendrobium blumei**
Onychium gracile	**Dendrobium gracile**
Onychium japonicum	**Dendrobium moniliforme**
Onychium lamellatum	**Dendrobium lamellatum**
Onychium mutabile	**Dendrobium mutabile**
Onychium nudum	**Dendrobium nudum**
Onychium tenellum	**Dendrobium tenellum**
Onychium tricuspe	**Dendrobium tricuspe**
Ophrys bivalvata	**Disa bivalvata**
Ophrys patens	**Disa tenuifolia**
Orchis cornuta	**Disa cornuta**
Orchis draconis	**Disa draconis**
Orchis filicornis	**Disa filicornis**
Orchis sagittalis	**Disa sagittalis**
Orchis tenella	**Disa tenella**
Orchis tenuifolia	**Disa tenella**
Orchis tripetaloides	**Disa tripetaloides**
Ormostema purpurea	**Dendrobium moniliforme**
Ormostemam albiflora	**Dendrobium moniliforme**
Orthopenthea atricapilla	**Disa atricapilla**
Orthopenthea bivalvata	**Disa bivalvata**

*For explanation see page 4, point 6
*Voir les explications page 9, point 6
*Para mayor explicación, véase la página 15, point 6

ALL NAMES

Orthopenthea bodkinii
Orthopenthea elegans
Orthopenthea fasciata
Orthopenthea minor
Orthopenthea obtusa
Orthopenthea richardiana
Orthopenthea rosea
Orthopenthea schizodioides
Orthopenthea telipogonis
Orthopenthea triloba
Oxystophyllum macrostoma
Pachyrhizanthe aberrans
Pachyrhizanthe aphyllum
Pachyrhizanthe macrorhizon
Pachyrhizanthe nipponicum
Pachyrhizanthe sagamiense
Pedilonum adolphi
Pedilonum aegle
Pedilonum aemulans
Pedilonum alderwereltianum
Pedilonum angustiflorum
Pedilonum anthrene
Pedilonum apertum
Pedilonum aphanochilum
Pedilonum asperifolium
Pedilonum auroroseum
Pedilonum begoniicarpum
Pedilonum biflorum
Pedilonum brachyphyta
Pedilonum bracteosum
Pedilonum brevicaule
Pedilonum brevilabium
Pedilonum brinchangense
Pedilonum bursigerum
Pedilonum caespitificum

Pedilonum calcarium

Pedilonum calicopis
Pedilonum caliculimentum
Pedilonum capituliflorum
Pedilonum cerasinum
Pedilonum chlorinum

Pedilonum chrysoglossum
Pedilonum chrysornis
Pedilonum coccinellum
Pedilonum cochleatum
Pedilonum concavissimum
Pedilonum cornutum
Pedilonum crenatifolium
Pedilonum crocatum
Pedilonum cuthbertsonii
Pedilonum cyanocentrum
Pedilonum cyperifolium

ACCEPTED NAMES

Disa bodkinii
Disa elegans
Disa fasciata
Disa minor
Disa richardiana
Disa richardiana
Disa rosea
Disa schizodioides
Disa telipogonis
Disa oligantha
Dendrobium aloifolium
Cymbidium macrorhizon
Cymbidium macrorhizon
Cymbidium macrorhizon
Cymbidium macrorhizon
Cymbidium macrorhizon
Dendrobium puniceum
Dendrobium erosum
Dendrobium erosum
Dendrobium alderwereltianum
Dendrobium angustiflorum
Dendrobium anthrene
Dendrobium apertum
Dendrobium aphanochilum
Dendrobium cuthbertsonii
Dendrobium nudum
Dendrobium subacaule
Dendrobium gemellum
Dendrobium vexillarius var. uncinatum
Dendrobium bracteosum
Dendrobium brevicaule
Dendrobium brevilabium
Dendrobium hasseltii
Dendrobium secundum
Dendrobium masarangense var.
   theionanthum
Dendrobium brevicaule subsp.
   calcarium
Dendrobium calicopis
Dendrobium caliculimentum
Dendrobium capituliflorum
Dendrobium puniceum
Dendrobium masarangense subsp.
   chlorinum
Dendrobium obtusum
Dendrobium dekockii
Dendrobium cuthbertsonii
Dendrobium cochleatum
Dendrobium obtusum
Dendrobium hasseltii
Dendrobium crenatifolium
Dendrobium crocatum
Dendrobium cuthbertsonii
Dendrobium cyanocentrum
Dendrobium violaceum subsp.
   cyperifolium

*For explanation see page 4, point 6
*Voir les explications page 9, point 6
*Para mayor explicación, véase la página 15, point 6

77

Part I: All Names / Tous les Noms / Todos los Nombres

ALL NAMES	ACCEPTED NAMES
Pedilonum dekockii	**Dendrobium dekockii**
Pedilonum delicatulum	**Dendrobium delicatulum**
Pedilonum derryi	**Dendrobium calicopis**
Pedilonum dichaeoides	**Dendrobium dichaeoides**
Pedilonum dichroma	**Dendrobium dichroma**
Pedilonum discrepans	**Dendrobium puniceum**
Pedilonum dryadum	**Dendrobium violaceum**
Pedilonum eitapense	**Dendrobium bracteosum**
Pedilonum eoum	**Dendrobium cumulatum**
Pedilonum erosum	**Dendrobium erosum**
Pedilonum euphues	**Dendrobium cuthbertsonii**
Pedilonum flammula	**Dendrobium flammula**
Pedilonum flavispiculum	**Dendrobium cyanocentrum**
Pedilonum fornicatum	**Dendrobium obtusum**
Pedilonum frigidum	**Dendrobium masarangense** var. **theionanthum**
Pedilonum fulgidum	**Dendrobium fulgidum**
Pedilonum geluanum	**Dendrobium hellwigianum**
Pedilonum geminiflorum	**Dendrobium violaceum**
Pedilonum gemma	**Dendrobium masarangense** var. **theionanthum**
Pedilonum glossotis	**Dendrobium catillare**
Pedilonum gnomus	**Dendrobium gnomus**
Pedilonum goldschmidtianum	**Dendrobium goldschmidtianum**
Pedilonum hasseltii	**Dendrobium hasseltii**
Pedilonum hellwigianum	**Dendrobium hellwigianum**
Pedilonum hollrungii	**Dendrobium smillieae**
Pedilonum hymenopterum	**Dendrobium hymenopterum**
Pedilonum inopinatum	**Dendrobium erosum**
Pedilonum jacobsonii	**Dendrobium jacobsonii**
Pedilonum junzaingense	**Dendrobium subacaule**
Pedilonum keysseri	**Dendrobium nebularum**
Pedilonum kuhlii	**Dendrobium hasseltii**
Pedilonum kuhlii	**Dendrobium thrysodes**
Pedilonum lamellatum	**Dendrobium lamellatum**
Pedilonum lancilabium	**Dendrobium lancilabium**
Pedilonum lapeyrouseoides	**Dendrobium cyanocentrum**
Pedilonum lateriflorum	**Dendrobium puniceum**
Pedilonum lawesii	**Dendrobium lawesii**
Pedilonum leucochysum	**Dendrobium bracteosum**
Pedilonum loesenerianum	**Dendrobium loesenerianum**
Pedilonum lompobatangense	**Dendrobium caliculimentum**
Pedilonum longicalcaratum	**Dendrobium chameleon**
Pedilonum maboroense	**Dendrobium undatialatum**
Pedilonum macrogenion	**Dendrobium macrogenion**
Pedilonum malvicolor	**Dendrobium malvicolor**
Pedilonum melinanthum	**Dendrobium melinanthum**
Pedilonum microblepharum	**Dendrobium vexillarius** var. **microblepharum**
Pedilonum mimiense	**Dendrobium constrictum**
Pedilonum minutum	**Dendrobium delicatulum**
Pedilonum mitriferum	**Dendrobium subclausum**
Pedilonum miyakei	**Dendrobium goldschmidtianum**
Pedilonum mohlianum	**Dendrobium mohlianum**
Pedilonum molle	**Dendrobium molle**
Pedilonum montigenum	**Dendrobium dekockii**

*For explanation see page 4, point 6
*Voir les explications page 9, point 6
*Para mayor explicación, véase la página 15, point 6

78

ALL NAMES

Pedilonum murkelense
Pedilonum nardoides
Pedilonum nebularum
Pedilonum neoebudanum
Pedilonum nubigenum
Pedilonum obtusum
Pedilonum occultum
Pedilonum oligoblepharon
Pedilonum ophioglossum
Pedilonum oreocharis
Pedilonum oreodoxa
Pedilonum oreogenum
Pedilonum pachyceras
Pedilonum panduriferum
Pedilonum pentagonum

Pedilonum pentapterum
Pedilonum petiolatum
Pedilonum phlox
Pedilonum pityphyllum
Pedilonum pumilio

Pedilonum puniceum
Pedilonum purpureum
Pedilonum putnamii
Pedilonum pyropum
Pedilonum quadrialatum
Pedilonum quinquecostatum
Pedilonum rarum
Pedilonum reinwardtii
Pedilonum retroflexum

Pedilonum rhaphiotes
Pedilonum rhodobotrys
Pedilonum rindjaniense
Pedilonum roseipes
Pedilonum rupestre
Pedilonum salmoneum
Pedilonum sanguinolentum
Pedilonum saruwagedicum

Pedilonum scarlatinum
Pedilonum scotiiferum

Pedilonum secundum
Pedilonum separatum
Pedilonum seranicum
Pedilonum serpens
Pedilonum smillieae
Pedilonum sophronites
Pedilonum strictum
Pedilonum subacaule
Pedilonum subclausum
Pedilonum subuliferum
Pedilonum sulphureum
Pedilonum talaudense

ACCEPTED NAMES

Dendrobium nebularum
Dendrobium nardoides
Dendrobium nebularum
Dendrobium mohlianum
Dendrobium nubigenum
Dendrobium obtusum
Dendrobium laevifolium
Dendrobium nardoides
Dendrobium smillieae
Dendrobium subacaule
Dendrobium oreodoxa
Dendrobium trichostomum
Dendrobium smillieae
Dendrobium panduriferum
Dendrobium brevicaule subsp.
    pentagonum
Dendrobium pentapterum
Dendrobium petiolatum
Dendrobium subclausum var. **phlox**
Dendrobium violaceum
Dendrobium masarangense subsp.
    masarangense
Dendrobium puniceum
Dendrobium purpureum
Dendrobium putnamii
Dendrobium crocatum
Dendrobium bracteosum
Dendrobium violaceum
Dendrobium rarum
Dendrobium purpureum
Dendrobium vexillarius var.
    retroflexum
Dendrobium hellwigianum
Dendrobium obtusum
Dendrobium rindjaniense
Dendrobium roseipes
Dendrobium rupestre
Dendrobium salmoneum
Dendrobium sanguinolentum
Dendrobium brevicaule subsp.
    pentagonum
Dendrobium puniceum
Dendrobium violaceum subsp.
    cyperifolium
Dendrobium secundum
Dendrobium calcaratum
Dendrobium seranicum
Dendrobium panduriferum
Dendrobium smillieae
Dendrobium cuthbertsonii
Dendrobium subclausum
Dendrobium subacaule
Dendrobium subclausum
Dendrobium subuliferum
Dendrobium sulphureum
Dendrobium purpureum subsp.

*For explanation see page 4, point 6
*Voir les explications page 9, point 6
*Para mayor explicación, véase la página 15, point 6

79

ALL NAMES	ACCEPTED NAMES
	candidulum
Pedilonum tenuicalcar	**Dendrobium violaceum**
Pedilonum theionanthum	**Dendrobium masarangense** var. theionanthum
Pedilonum trachyphyllum	**Dendrobium cuthbertsonii**
Pedilonum trialatum	**Dendrobium vexillarius** var. **uncinatum**
Pedilonum trichostomum	**Dendrobium trichostomum**
Pedilonum tricostatum	**Dendrobium subacaule**
Pedilonum triviale	**Dendrobium calcaratum**
Pedilonum tumidulum	**Dendrobium nebulare**
Pedilonum uncinatum	**Dendrobium vexillarius** var. **uncinatum**
Pedilonum undatialatum	**Dendrobium undatialatum**
Pedilonum undulatum	**Dendrobium hymenophyllum**
Pedilonum verruculosum	**Dendrobium verruculosum**
Pedilonum vexillarius	**Dendrobium vexillarius**
Pedilonum victoriae-reginae	**Dendrobium victoriae-reginae**
Pedilonum violaceum	**Dendrobium violaceum**
Pedilonum vitellinum	**Dendrobium mohlianum**
Pedilonum woluense	**Dendrobium woluense**
Penthea atricapilla	**Disa atricapilla**
Penthea elegans	**Disa elegans**
Penthea filicornis	**Disa filicornis**
Penthea melaleuca	**Disa bivalvata**
Penthea minor	**Disa minor**
Penthea obtusa	**Disa richardiana**
Penthea patens	**Disa tenuifolia**
Penthea reflexa	**Disa filicornis**
Penthea triloba	**Disa oligantha**
Phaedrosanthus cochleatus	**Encyclia cochleata**
Phyllorchis lichenastrum	**Dendrobium lichenastrum**
Prosthechea glauca	**Encyclia glauca**
Sarcopodium parvulum	**Dendrobium delicatulum** subsp. parvulum
Sarcopodium prasinum	**Dendrobium prasinum**
Satyrium calceatum	**Disa buchenaviana**
Satyrium cornutum	**Disa cornuta**
Satyrium cylindrica	**Disa cylindrica**
Satyrium draconis	**Disa draconis**
Satyrium excelsum	**Disa tripetaloides**
Satyrium ferrugineum	**Disa ferruginea**
Satyrium grandiflora	**Disa uniflora**
Satyrium sagittale	**Disa sagittalis**
Satyrium secundatum	**Disa racemosa**
Satyrium tenellum	**Disa tenella**
Sayeria aberrans	**Dendrobium aberrans**
Sayeria acutisepala	**Dendrobium acutisepalum**
Sayeria alexandrae	**Dendrobium alexandrae**
Sayeria amphigenya	**Dendrobium amphigenyum**
Sayeria armeniaca	**Dendrobium armeniacum**
Sayeria atroviolacea	**Dendrobium atroviolaceum**
Sayeria bairdiana	**Dendrobium fellowsii**
Sayeria bifalcis	**Dendrobium bifalce**
Sayeria bilocularis	**Dendrobium biloculare**
Sayeria convoluta	**Dendrobium convolutum**
Sayeria curvimenta	**Dendrobium curvimentum**
Sayeria dendrocolloides	**Dendrobium dendrocolloides**

*For explanation see page 4, point 6
*Voir les explications page 9, point 6
*Para mayor explicación, véase la página 15, point 6

ALL NAMES	ACCEPTED NAMES
Sayeria diceras	Dendrobium diceras
Sayeria euryantha	Dendrobium euryanthum
Sayeria eustachya	Dendrobium forbesii
Sayeria eximia	Dendrobium eximium
Sayeria finisterrae	Dendrobium finisterrae
Sayeria forbesii	Dendrobium forbesii
Sayeria hodgkinsonii	Dendrobium hodgkinsonii
Sayeria incurvilabia	Dendrobium dendrocolloides
Sayeria informis	Dendrobium informe
Sayeria johnsoniae	Dendrobium johnsoniae
Sayeria laurensii	Dendrobium laurensii
Sayeria leucohybos	Dendrobium leucohybos
Sayeria macrophylla	Dendrobium macrophyllum
Sayeria mayandyi	Dendrobium mayandyi
Sayeria minutiflora	Dendrobium minutiflorum
Sayeria mooreana	Dendrobium mooreanum
Sayeria muscifera	Dendrobium macrophyllum
Sayeria otaguroana	Dendrobium otaguroanum
Sayeria pachystele	Dendrobium pachystele
Sayeria paradoxa	Dendrobium cruttwellii
Sayeria pleurodes	Dendrobium pleurodes
Sayeria polysema	Dendrobium polysema
Sayeria pseudotokai	Dendrobium macranthum
Sayeria punamensis	Dendrobium punamense
Sayeria rhodosticta	Dendrobium rhodostictum
Sayeria rhomboglossa	Dendrobium rhomboglossum
Sayeria rigidifolia	Dendrobium rigidifolium
Sayeria ruginosa	Dendrobium ruginosum
Sayeria ruttenii	Dendrobium ruttenii
Sayeria simplex	Dendrobium simplex
Sayeria spectabilis	Dendrobium spectabile
Sayeria subquadrata	Dendrobium subquadratum
Sayeria terrestris	Dendrobium terrestre
Sayeria torricellensis	Dendrobium torricellense
Sayeria uncipes	Dendrobium uncipes
Sayeria violascens	Dendrobium violascens
Sayeria wisselensis	Dendrobium wisselense
Sayeria woodsii	Dendrobium woodsii
Schizodium maculatum	Disa maculata
Serapias melaleuca	Disa bivalvata
Serapias patens	Disa tenuifolia
Sobralia citrina	Encyclia citrina
Thelychiton brachypus	Dendrobium brachypus
Tropilis adae	Dendrobium adae
Tropilis aemula	Dendrobium aemulum
Tropilis comptonii	Dendrobium comptonii
Tropilis × delicata	Dendrobium × delicatum
Tropilis drake-castilloi	Dendrobium comptonii
Tropilis falcorostra	Dendrobium falcorostrum
Tropilis fleckeri	Dendrobium fleckeri
Tropilis gracilicaulis	Dendrobium gracilicaule
Tropilis × gracillimum	Dendrobium × gracillimum
Tropilis kingiana	Dendrobium kingianum
Tropilis moorei	Dendrobium moorei
Tropilis ruppiana	Dendrobium jonesii
Tropilis speciosa	Dendrobium speciosum

*For explanation see page 4, point 6
*Voir les explications page 9, point 6
*Para mayor explicación, véase la página 15, point 6

ALL NAMES	ACCEPTED NAMES
Tropilis × suffusa	**Dendrobium × suffusum**
Tropilis tetragona	**Dendrobium tetragonum**
Yoania aberrans	**Cymbidium macrorhizon**

*For explanation see page 4, point 6
*Voir les explications page 9, point 6
*Para mayor explicación, véase la página 15, point 6

82

PART II: ORCHIDACEAE BINOMIALS IN CURRENT USAGE
Ordered alphabetically on Accepted Names for the genera:

Cymbidium, Dendrobium (selected sections only*), Disa, Dracula* and *Encyclia*

Deuxième partie: BINOMES D'ORCHIDACEAE ACTUELLEMENT EN USAGE
Par ordre alphabétique des noms reconnus pour les genre:

Cymbidium, Dendrobium (certaines sections sélectionnées), *Disa, Dracula* et *Encyclia*

Parte II: ORCHIDACEAE BINOMIALES UTILIZADOS NORMALMENTE
Presentados por orden alfabético: nombres aceptados para el genero:

Cymbidium, Dendrobium (únicamente las secciones seleccionadas), *Disa, Dracula* y *Encyclia*

Part II: Cymbidium

CYMBIDIUM BINOMIALS IN CURRENT USAGE

CYMBIDIUM BINOMES ACTUELLEMENT EN USAGE

CYMBIDIUM BINOMIALES UTILIZADOS NORMALMENTE

Cymbidium aloifolium (L.) Sw.
Aerides borassii Buch. Ham. ex Sm.
Cymbidium atropurpureum sensu T.K.Yen
Cymbidium erectum Wight
Cymbidium intermedium H.G.Jones
Cymbidium pendulum (Roxb.) Sw.
Cymbidium simulans Rolfe
Epidendrum aloides Curtis
Epidendrum aloifolium L.
Epidendrum pendulum Roxb.

Distribution: Bangladesh, Cambodia, China, India, Indonesia, Lao People's Democratic Republic (the), Malaysia, Myanmar, Nepal, Sri Lanka, Thailand, Viet Nam

Cymbidium atropurpureum (Lindl.) Rolfe
Cymbidium atropurpureum var. *olivaceum* J.J.Sm.
Cymbidium finlaysonianum var. *atropurpureum* (Lindl.) J.H.Veitch
Cymbidium pendulum var. *atropurpureum* Lindl.
Cymbidium pendulum var. *purpureum* W.Watson

Distribution: Indonesia, Malaysia, Philippines (the), Thailand, Viet Nam

Cymbidium bananense Gagnep.

Distribution: Viet Nam

Cymbidium bicolor Lindl.
Cymbidium aloifolium sensu Jayaw.

Distribution: Bhutan, Cambodia, China, India, Indonesia, Lao People's Democratic Republic, Malaysia, Myanmar, Nepal, Philippines (the), Sri Lanka, Thailand, Viet Nam

Cymbidium bicolor subsp. **bicolor**

Distribution: India, Sri Lanka

Cymbidium bicolor subsp. **obtusum** Du Puy & P.J.Cribb
Cymbidium bicolor sensu Seidenf.
Cymbidium crassifolium Wall.
Cymbidium flaccidum Schltr.
Cymbidium mannii Rchb.f.
Cymbidium pendulum sensu Duthie
Cymbidium pendulum sensu H.Hara, Stearn & Williams
Cymbidium pendulum sensu King. & Pantl.

Cymbidium pendulum sensu Pradhan
Cymbidium pendulum sensu Schltr.
Cymbidium pendulum sensu Y.S.Wu & S.C.Chen

Distribution: Bhutan, Cambodia, China, India, Lao People's Democratic Republic, Myanmar, Nepal, Thailand, Viet Nam

Cymbidium bicolor subsp. **pubescens** (Lindl.) Du Puy & P.J.Cribb
Cymbidium aloifolium sensu Blume
Cymbidium aloifolium var. *pubescens* (Lindl.) Ridl.
Cymbidium celebicum (Schltr.) Schltr.
Cymbidium pubescens Lindl.
Cymbidium pubescens var. *celebicum* Schltr.

Distribution: Indonesia, Malaysia, Philippines (the)

Cymbidium borneense J.J.Wood

Distribution: Malaysia

Cymbidium canaliculatum R.Br.
Cymbidium canaliculatum var. *barrettii* Nicholls
Cymbidium canaliculatum var. *marginatum* Rupp
Cymbidium canaliculatum var. *sparkesii* (Rendle) F.M.Bailey
Cymbidium hilii F.Muell.
Cymbidium sparkesii Rendle

Distribution: Australia

Cymbidium chloranthum Lindl.
Cymbidium pulchellum Schltr.
Cymbidium sanguineolentum Teijsm. & Binn.
Cymbidium sanguineum Teijsm. & Binn.
Cymbidium variciferum Rchb.f.

Distribution: Indonesia, Malaysia

Cymbidium cochleare Lindl.
Cymbidium babae (Kudo ex Masam.) Masam.
Cymbidium kanran var. *babae* (Kudo) S.S.Ying
Cyperorchis babae Kudo ex Masam.
Cyperorchis cochlearis (Lindl.) Benth.

Distribution: China, China (Taiwan), India, Myanmar, Thailand

Cymbidium cyperifolium Wall. ex Lindl.
Cymbidium aliciae Quisumb.
Cymbidium carnosum Griff.
Cymbidium viridiflorum Griff.
Cyperorchis wallichii Blume

Distribution: Bhutan, Cambodia, China, India, Myanmar, Nepal, Philippines (the), Thailand, Viet Nam

Cymbidium cyperifolium subsp. **arrogans** Du Puy & P.J.Cribb

Distribution: Cambodia, Myanmar, Philippines (the), Thailand

Cymbidium cyperifolium subsp. **cyperifolium**

Distribution: Bhutan, China, India, Nepal, Viet Nam

Cymbidium dayanum Rchb.f.
Cymbidium acutum Ridl.
Cymbidium alborubens Makino
Cymbidium angustifolium Ames & C.Schweinf.
Cymbidium dayanum var. *austro-japonicum* Tuyama
Cymbidium eburneum var. *austro-japonicum* (Tuyama) M.Hiroe
Cymbidium eburneum var. *dayana* Hook.f.
Cymbidium leachianum Rchb.f.
Cymbidium pendulum Ridl.
Cymbidium poilanei Gagnep.
Cymbidium pulcherrimum Sander
Cymbidium simonsianum King & Pantl.
Cymbidium sutepense Rolfe ex Downie

Distribution: Cambodia, China, China (Taiwan), India, Indonesia, Japan, Lao People's Democratic Republic, Malaysia, Philippines (the), Thailand, Viet Nam

Cymbidium defoliatum Y.S.Wu & S.C.Chen

Distribution: China

Cymbidium devonianum Paxton
Cymbidium sikkimense Hook.f.

Distribution: Bhutan, India, Nepal, Thailand, Viet Nam

Cymbidium eburneum Lindl.
Cymbidium eburneum var. *dayi* A.Jenn.
Cymbidium eburneum var. *obtusum* Rchb.f.
Cymbidium eburneum var. *philbrickianum* Rchb.f.
Cymbidium syringodorum Griff.
Cyperorchis eburnea (Lindl.) Schltr.

Distribution: China, India, Myanmar, Nepal

Cymbidium elegans Lindl.
Arethusantha bletioides Finet

Cymbidium densiflorum Griff.
Cymbidium elegans var. *lutescens* Hook.f.
Cymbidium elegans var. *obcordatum* Rchb.f.
Cymbidium longifolium D.Don
Cyperorchis elegans (Lindl.) Blume
Cyperorchis elegans var. *blumei* Hort.
Grammatophyllum elegans (Lindl.) Rchb.f.

Distribution: Bhutan, China, India, Myanmar, Nepal

Cymbidium elongatum J.J.Wood, Du Puy & Shim

Distribution: Malaysia

Cymbidium ensifolium (L.) Sw.
Cymbidium albo-marginatum Makino
Cymbidium arrogans Hayata
Cymbidium ecristatum Steud.
Cymbidium ensifolium var. *arrogans* (Hayata) T.S.Liu & H.J.Su
Cymbidium ensifolium var. *arrogans* T.K.Yen
Cymbidium ensifolium var. *estriatum* Lindl.
Cymbidium ensifolium var. *misericors* (Hayata) T.P.Lin
Cymbidium ensifolium var. *striatum* Lindl.
Cymbidium ensifolium var. *yakibaran* (Makino) Y.S.Wu & S.C.Chen
Cymbidium gonzalesii Quisumb.
Cymbidium gyokuchin Makino
Cymbidium gyokuchin var. *soshin* Makino
Cymbidium gyokuchin var. *arrogans* (Hayata) S.S.Ying
Cymbidium kanran var. *misericors* (Hayata) S.S.Ying
Cymbidium koran Makino
Cymbidium micans Schauer
Cymbidium misericors Hayata
Cymbidium niveo-marginatum Makino
Cymbidium rubrigemmum Hayata
Cymbidium shimaran Makino
Cymbidium xiphiifolium Lindl.
Cymbidium yakibaran Makino
Epidendrum ensifolium L.
Epidendrum sinense Redoute
Jensoa ensata (Thunb.) Raf.
Limodorum ensatum Thunb.

Distribution: Cambodia, China, China (Taiwan), India, Indonesia, Japan, Lao People's
Democratic Republic, Malaysia, Papua New Guinea, Philippines (the), Sri Lanka, Thailand, Viet
Nam

Cymbidium ensifolium subsp. **ensifolium**

Distribution: China, China (Taiwan), Japan, Philippines (the), Thailand, Viet Nam

Cymbidium ensifolium subsp. **haematodes** (Lindl.) Du Puy & P.J.Cribb
Cymbidium acuminatum M.A.Clem. & D.L.Jones

Part II: Cymbidium

Cymbidium ensifolium sensu Du Puy & A.L.Lamb
Cymbidium ensifolium sensu Holttum
Cymbidium ensifolium sensu J.B.Comber
Cymbidium ensifolium sensu J.J.Sm.
Cymbidium ensifolium sensu J.J.Wood
Cymbidium ensifolium sensu Ridl.
Cymbidium ensifolium sensu Seidenf.
Cymbidium ensifolium var. *haematodes* (Lindl.) Trimen
Cymbidium siamense Rolfe ex Downie
Cymbidium sundaicum Schltr.
Cymbidium sundaicum var. *estriata* Schltr.

Distribution: India, Indonesia, Malaysia, Papua New Guinea, Sri Lanka, Thailand

Cymbidium erythraeum Lindl.
Cymbidium longifolium sensu Lindl.
Cymbidium hennisianum Schltr.
Cyperorchis longifolia (D. Don) Schltr.
Cyperorchis hennisiana (Schltr.) Schltr.

Distribution: China, Myanmar, Thailand

Cymbidium erythrostylum Rolfe
Cymbidium erythrostylum var. *magnificum* Hort.
Cyperorchis erythrostyla (Rolfe) Schltr.

Distribution: Viet Nam

Cymbidium faberi Rolfe
Cymbidium cerinum Schltr.
Cymbidium fukienense T.K.Yen
Cymbidium oiwakensis Hayata
Cymbidium scabroserrulatum Makino

Distribution: China, China (Taiwan), India, Nepal

Cymbidium faberi var. **faberi**

Distribution: China, China (Taiwan)

Cymbidium faberi var. **szechuanicum** (Y.S.Wu & S.C.Chen) Y.S.Wu & S.C.Chen
Cymbidium cyperifolium sensu auct. non Wall. ex Lindl.
Cymbidium szechuanicum Y.S.Wu & S.C.Chen

Distribution: China, India, Nepal

Cymbidium finlaysonianum Lindl.
Cymbidium aloifolium sensu Guillaumin
Cymbidium pendulum sensu Blume
Cymbidium pendulum sensu Lindl.

Cymbidium pendulum sensu S.Vidal
Cymbidium pendulum var. *brevilabre* Lindl.
Cymbidium tricolor Miq.
Cymbidium wallichii Lindl.

Distribution: Cambodia, Indonesia, Malaysia, Philippines (the), Thailand, Viet Nam

Cymbidium floribundum Lindl.
Cymbidium floribundum var. *pumilum* (Rolfe) Y.S.Wu & S.C.Chen
Cymbidium illiberale Hayata
Cymbidium pumilum Rolfe

Distribution: China, China (Taiwan), Viet Nam

Cymbidium goeringii (Rchb.f.) Rchb.f.
Cymbidium chuen-lan C.Chow
Cymbidium formosanum Hayata
Cymbidium forrestii Rolfe
Cymbidium mackinnoni Duthie
Cymbidium pseudovirens Schltr.
Cymbidium tentyozanense Masam.
Cymbidium uniflorum T.K.Yen
Cymbidium virens Rchb.f.
Cymbidium virescens sensu Lindl.
Cymbidium yunnanense Schltr.
Maxillaria goeringii Rchb.f.

Distribution: China, China (Taiwan), India, Japan, Korea (the Republic of)

Cymbidium goeringii var. **goeringii**

Distribution: China, China (Taiwan), India, Japan, Korea (the Republic of)

Cymbidium goeringii var. **longibracteateatum** Y.S.Wu & S.C.Chen
Cymbidium longibracteateatum Y.S.Wu & S.C.Chen

Distribution: China

Cymbidium goeringii var. **serratum** (Schltr.) Y.S.Wu & S.C.Chen
Cymbidium formosanum var. *gracillimum* (Fukuy.) Tang S.Liu & H.J.Su
Cymbidium goeringii var. *arrogans* F.Maek.
Cymbidium gracillimum Fukuy.
Cymbidium serratum Schltr.

Distribution: China, China (Taiwan), Japan

Cymbidium goeringii var. **tortisepalum** (Fukuy.) Y.S.Wu & S.C.Chen
Cymbidium tortisepalum Fukuy.
Cymbidium tortisepalum var. *viridiflorum* S.S.Ying
Cymbidium tsukengensis C.Chow

Part II: Cymbidium

Distribution: China, China (Taiwan)

Cymbidium gongshanense H. Li & G.H. Feng

Distribution: China

Cymbidium hartinahianum J.B.Comber & Nasution

Distribution: Indonesia

Cymbidium hookerianum Rchb.f.
Cymbidium giganteum var. *hookerianum* (Rchb.f.) Bois
Cymbidium grandiflorum Griff.
Cymbidium grandiflorum var. *punctatum* Cogn.
Cyperorchis grandiflora (Griff.) Schltr.

Distribution: Bhutan, China, India, Nepal, Viet Nam

Cymbidium insigne Rolfe
Cymbidium insigne var. *album* Hort.
Cymbidium insigne var. *sanderi* (O'Brien) Hort.
Cymbidium sanderi O'Brien
Cyperorchis insignis (Rolfe) Schltr.

Distribution: China, Thailand, Viet Nam

Cymbidium iridioides D.Don
Cymbidium giganticum Wall. ex Lindl.

Distribution: Bhutan, China, India, Myanmar, Nepal, Viet Nam

Cymbidium kanran Makino
Cymbidium faberi var. *omeiense* (Y.S.Wu & S.C.Chen) Y.S.Wu & S.C.Chen
Cymbidium kanran var. *latifolium* Makino
Cymbidium linearisepalum Yamam.
Cymbidium misericors var. *oreophyllum* (Hayata) Hayata
Cymbidium omeiense Y.S.Wu & S.C.Chen
Cymbidium oreophyllum Hayata
Cymbidium purpureo-hiemale Hayata
Cymbidium sinokanran T.K.Yen
Cymbidium sinokanran var. *atropurpureum* T.K.Yen
Cymbidium tentyozanense T.P.Lin
Cymbidium tosyaense Masam.

Distribution: China, China (Taiwan), Japan, Korea (the Republic of)

Cymbidium kinabaluense K.M. Wong & C.L. Chan

Distribution: Malaysia

Cymbidium lancifolium Hook.
Cymbidium aspidistrifolium Fukuy.
Cymbidium bambusifolium Fowlie, Mark & H.S.Ho
Cymbidium caulescens Ridl.
Cymbidium cuspidatum Blume
Cymbidium gibsonii Lindl.
Cymbidium javanicum Blume
Cymbidium javanicum var. *aspidistrifolium* (Fukuy.) F.Maek.
Cymbidium javanicum var. *pantlingii* F.Maek.
Cymbidium kerrii Rolfe
Cymbidium lancifolium var. *aspidistrifolium* (Fukuy.) S.S.Ying
Cymbidium lancifolium var. *syunitianum* (Fukuy.) S.S.Ying
Cymbidium maclehoseae S.Y.Hu
Cymbidium nagifolium Masam.
Cymbidium papuanum Schltr.
Cymbidium robustum Gilli
Cymbidium syunitianum Fukuy.

Distribution: Bhutan, China, China (Taiwan), India, Indonesia, Japan, Lao People's Democratic Republic, Malaysia, Myanmar, Nepal, Papua New Guinea, Thailand, Viet Nam

Cymbidium lowianum (Rchb.f.) Rchb.f.
Cymbidium giganteum var. *lowianum* Rchb.f.
Cymbidium hookerianum var. *lowianum* (Rchb.f.) Y.S.Wu & S.C.Chen
Cymbidium lowianum var. *concolor* Rolfe
Cymbidium lowianum var. *flaveolum* L.Linden
Cymbidium lowianum var. *superbissimum* L.Linden
Cymbidium lowianum var. *viride* R.Warner & N.H.Williams
Cyperorchis lowiana (Rchb.f.) Schltr.

Distribution: China, Myanmar, Thailand, Viet Nam

Cymbidium lowianum var. **iansonii** (Rolfe) P.J.Cribb & Du Puy
Cymbidium grandiflorum var. *kalawensis* Colyear

Distribution: China, Myanmar

Cymbidium lowianum var. **lowianum**

Distribution: China, Myanmar, Thailand, Viet Nam

Cymbidium macrorhizon Lindl.
Aphyllorchis aberrans (Finet) Schltr.
Bletia nipponica Franch. & Sav.
Cymbidium aberrans (Finet) Schltr.
Cymbidium aphyllum Ames & Schltr.
Cymbidium nipponicum (Franch. & Sav.) Rolfe
Cymbidium pedicellatum Finet
Cymbidium sagamiense (Nakai) Makino & Nemoto

Part II: Cymbidium

Cymbidium szechuanensis S.Y.Hu
Pachyrhizanthe aberrans (Finet) Nakai
Pachyrhizanthe aphyllum (Ames & Schltr.) Nakai
Pachyrhizanthe macrorhizon (Lindl.) Nakai
Pachyrhizanthe nipponicum (Franch. & Sav.) Nakai
Pachyrhizanthe sagamiense Nakai
Yoania aberrans Finet

Distribution: China, China (Taiwan), India, Japan, Lao People's Democratic Republic (the), Myanmar, Nepal, Pakistan, Thailand, Viet Nam

Cymbidium madidum Lindl.
Cymbidium albuciflorum F.Muell.
Cymbidium iridifolium A.Cunn.
Cymbidium leai Rendle
Cymbidium leroyi St.Cloud
Cymbidium pulchellum F.Muell.
Cymbidium pulchellum var. *leroyi* (St.Cloud) Menninger
Cymbidium queeneanum Klinge

Distribution: Australia

Cymbidium maguanense F.Y.Liu

Distribition: China

Cymbidium mastersii Griff. ex Lindl.
Cymbidium affine Griff.
Cymbidium mastersii var. *album* Rchb.f.
Cymbidium micromeson Lindl.
Cyperorchis mastersii Benth.

Distribution: China, India, Myanmar, Thailand

Cymbidium munronianum King & Pantl.
Cymbidium ensifolium var. *munronianum* (King & Pantl.) T.Tang & F.T.Wang

Distribution: Bhutan, India

Cymbidium nanulum Y.S.Wu & S.C.Chen

Distribution: China

Cymbidium parishii Rchb.f.
Cymbidium eburneum var. *parishii* (Rchb.f.) Hook.f.
Cyperorchis parishii(Rchb.f.) Schltr.

Distribution: Myanmar

Cymbidium qiubeiense K.M.Feng & H.Li

Distribution: China

Cymbidium rectum Ridl.

Distribution: Malaysia

Cymbidium roseum J.J.Sm.
  Cyperorchis rosea (J.J.Sm.) Schltr.

Distribution: Indonesia, Malaysia

Cymbidium sanderae (Rolfe) P.J.Cribb & Du Puy
  Cymbidium pulchellum var. *sanderae* Rolfe

Distribution: Viet Nam

Cymbidium schroederi Rolfe
  Cyperorchis schroederi (Rolfe) Schltr.

Distribution: Viet Nam

Cymbidium sigmoideum J.J.Sm.
  Cyperorchis sigmoidea (J.J.Sm.) J.J.Sm.

Distribution: Indonesia, Malaysia

Cymbidium sinense (G.Jacks. & H.C.Andr.) De Wild.
  Cymbidium albo-jucundissimum Hayata
  Cymbidium chinense Heynh.
  Cymbidium ensifolium auct. non Sw. Hook.f.
  Cymbidium ensifolium var. *munronianum* sensu T.Tang & F.T.Wang
  Cymbidium fragrans Salisb.
  Cymbidium hoosai Makino
  Cymbidium sinense var. *albo-jucundissimum* (Hayata) Masam.
  Cymbidium sinense var. *album* T.K.Yen
  Cymbidium sinense var. *arrogans* Hayata
  Cymbidium sinense var. *bellum* T.K.Yen
  Epidendrum sinense G.Jacks ex H.C.Andr.

Distribution: China, China (Taiwan), India, Japan, Myanmar, Thailand, Viet Nam

Cymbidium suave R.Br.
  Cymbidium gomphocarpum Fitzg.

Distribution: Australia

Part II: Cymbidium

Cymbidium suavissimum Hort. ex C.H.Curtis

Distribution: China, Myanmar, Viet Nam

Cymbidium tracyanum L.Castle
  Cyperorchis traceyana (L.Castle) Schltr.

Distribution: China, Myanmar, Thailand

Cymbidium tigrinum Parish ex Hook.
  Cyperorchis tigrina (Parish ex Hook.) Schltr.

Distribution: China, India, Myanmar

Cymbidium wenshanense Y.S.Wu & F.Y.Liu

Distribution: China

Cymbidium whiteae King & Pantl.
  Cyperorchis whiteae (King & Pantl.) Schltr.

Distribution: India

Cymbidium wilsonii (Rolfe) Rolfe
  Cymbidium giganteum var. *willsonii* Rolfe
  Cyperorchis wilsonii (Rolfe) Schltr.

Distribution: China

DENDROBIUM BINOMIALS IN CURRENT USAGE
Selected sections only:

DENDROBIUM BINOMES ACTUELLEMENT EN USAGE
Certaines sections sélectionnées:

DENDROBIUM BINOMIALES UTILIZADOS NORMALMENTE
Únicamente las secciones seleccionada:

Aporum
Breviflores
Callista
Calyptrochilus
Dendrobium
Dendrocryne
Formosae
Fytchianthe
Latouria

Macrocladium
Oxyglossum
Pedilonum
Phalaenanthe
Rhizobium
Rhopalanthe
Spatulata
Stachyobium

Dendrobium aberrans Schltr.
 Sayeria aberrans (Schltr.) Rauschert

Distribution: Papua New Guinea

Dendrobium aciculare Lindl.
 Aporum aciculare (Lindl.) Rauschert
 Aporum koeteianum (Schltr.) Rauschert
 Callista acicularis (Lindl.) Kuntze
 Dendrobium koeteianum Schltr.

Distribution: Indonesia, Malaysia, Singapore, Thailand

Dendrobium acinaciforme Roxb.
 Aporum acinaciforme (Roxb.) Griff.
 Aporum banaense (Gagnep.) Rauschert
 Aporum serra auct. non Lindl.
 Callista acinaciformis (Roxb.) Kuntze
 Callista spatella (Rchb.f.) Kuntze
 Dendrobium banaense Gagnep.
 Dendrobium spatella Rchb.f.

Distribution: Cambodia, China, India, Lao People's Democratic Republic, Myanmar, Thailand, Viet Nam

Dendrobium acutimentum J.J.Sm.

Distribution: Indonesia

Dendrobium acutisepalum J.J.Sm.

95

Part II: Dendrobium

Sayeria acutisepala (J.J.Sm.) Rauschert

Distribution: Indonesia, Papua New Guinea

Dendrobium adae F.M.Bailey
Callista adae (F.M.Bailey) Kuntze
Dendrobium ancorarium Rupp
Dendrobium palmerstoniae Schltr.
Tropilis adae (F.M.Bailey) Butzin
Tropilis adae (F.M.Bailey) Rauschert

Distribution: Australia

Dendrobium aduncum Wall. ex Lindl.
Callista aduncum (Wall. ex Lindl.) Kuntze

Distribution: Bhutan, China, India, Myanmar, Thailand, Viet Nam

Dendrobium aemulum R.Br.
Callista aemula (R.Br.) Kuntze
Tropilis aemula (R.Br.) Raf.

Distribution: Australia

Dendrobium affine (Decne) Steud.
Callista dicupha (F.Muell.) Kuntze
Callista leucolophota (Rchb.f.) Kuntze
Dendrobium bigibbum sensu F.Muell.
Dendrobium dicuphum F.Muell.
Dendrobium dicuphum var. *album* Hort. ex E.C.Cooper
Dendrobium dicuphum var. *grandiflorum* Rupp & T.E.Hunt
Dendrobium leucolophotum Rchb.f.
Dendrobium urvillei Finet
Onychium affine Decne.

Distribution: Australia, Indonesia

Dendrobium alabense J.J.Wood

Distribution: Malaysia

Dendrobium alaticaulinum P.Royen

Distribution: Papua New Guinea

Dendrobium albayense Ames

Distribution: Philippines (the)

Dendrobium albosanguineum Lindl.
Callista albosanguinea (Lindl. & Paxt.) Kuntze

Distribution: Myanmar, Thailand

Dendrobium alboviride Hayata

Distribution: China (Taiwan)

Dendrobium alderwereltianum J.J.Sm.
Pedilonum alderwereltianum (J.J.Sm.) Rauschert

Distribution: Indonesia

Dendrobium alexandrae Schltr.
Latourorchis alexandrae (Schltr.) Brieger
Sayeria alexandrae (Schltr.) Rauschert

Distribution: Papua New Guinea

Dendrobium aloifolium (Blume) Rchb.f.
Aporum aloifolium (Blume) Brieger
Aporum cochinchinense (Ridl.) Brieger
Aporum micranthum Griff.
Aporum serra Lindl.
Callista aloefolia (Blume) Kuntze
Callista borneoensis Kuntze
Callista *micrantha* (Griff.) Kuntze
Dendrobium cochinchinense Ridl.
Dendrobium micranthum (Griff.) Lindl.
Dendrobium serra (Lindl.) Lindl.
Macrostomium aloefolium Blume
Oxystophyllum macrostoma Hassk.

Distribution: Cambodia, Indonesia, Lao People's Democratic Republic, Malaysia, Myanmar, Singapore, Thailand, Viet Nam

Dendrobium amabile (Lour.) O'Brien
Callista amabilis Lour.
Dendrobium bronckartii De Wild.
Epidendrum callista Raeusch

Distribution: Viet Nam

Dendrobium amblyogenium Schltr.
Pedilonum amblyogenium (Schltr.) Rauschert

Distribution: Indonesia

Part II: Dendrobium

Dendrobium amethystoglossum Rchb.f.
  Callista amethystoglossa (Rchb.f.) Kuntze

Distribution: Philippines (the)

Dendrobium amoenum Lindl.
  Callista amoena (Wall. ex Lindl.) Kuntze

Distribution: Bhutan, India, Myanmar, Nepal

Dendrobium amphigenyum Ridl.
  Dendrobium fantasticum L.O.Williams
  Sayeria amphigenya (Ridl.) Rauschert

Distribution: Indonesia, Papua New Guinea

Dendrobium anceps Sw.
  Aporum anceps (Sw.) Lindl.
  Callista anceps (Sw.) Kuntze

Distribution: Bhutan, India, Lao People's Democratic Republic, Myanmar, Nepal, Thailand

Dendrobium andreemillarae T.M.Reeve

Distribution: Papua New Guinea

Dendrobium angiense J.J.Sm.
  Pedilonum angiense (J.J.Sm.) Rauschert

Distribution: Indonesia

Dendrobium angustiflorum J.J.Sm.
  Pedilonum angustiflorum (J.J.Sm.) Rauschert

Distribution: Indonesia

Dendrobium annae J.J.Sm.

Distribution: Indonesia

Dendrobium annamense Rolfe

Distribution: Viet Nam

Dendrobium anosmum Lindl.
  Callista anosma (Lindl.) Kuntze

Callista macrophylla (Lindl.) Kuntze
Callista scortechinii (Hook.f.) Kuntze
Dendrobium leucorhodum Schltr.
Dendrobium macranthum Hook.
Dendrobium macrophyllum auct. non A.Rich
Dendrobium pulchellum auct. non Rchb.f.
Dendrobium scortechinii Hook.f.
Dendrobium superbum Rchb.f.

Distribution: India, Indonesia, Lao People's Democratic Republic, Malaysia, Myanmar, Papua New Guinea, Philippines (the), Sri Lanka, Thailand, Viet Nam

Dendrobium antennatum Lindl.
Callista antennata (Lindl.) Kuntze
Dendrobium d'albertisii Rchb.f.

Distribution: Australia, Indonesia, Papua New Guinea, Solomon Islands

Dendrobium anthrene Ridl.
Pedilonum anthrene (Ridl.) Rauschert

Distribution: Indonesia, Malaysia

Dendrobium apertum Schltr.
Pedilonum apertum (Schltr.) Rauschert

Distribution: Papua New Guinea

Dendrobium aphanochilum Kraenzl.
Pedilonumaphanochilum (Kraenzl.) Rauschert

Distribution: Indonesia

Dendrobium aphyllum (Roxb.) Fischer
Callista aphylla (Roxb.) Kuntze
Cymbidium aphyllum (Roxb.) Sw.
Dendrobium cucullatum R.Br.
Dendrobium evaginatum Gagnep.
Dendrobium madrasense Hawkes
Dendrobium oxyphyllum Gagnep.
Dendrobium pierardii (Roxb. ex Hook.) Fischer
Limodorum aphyllum Roxb.

Distribution: Bhutan, Cambodia, China, India, Lao People's Democratic Republic, Malaysia, Myanmar, Nepal, Thailand, Viet Nam

Dendrobium aqueum Lindl.
Callista aquea (Lindl.) Kuntze
Dendrobium album Wt.

Part II: Dendrobium

Distribution: India

Dendrobium arachnoglossum Rchb.f.

Distribution: Papua New Guinea

Dendrobium arcuatum J.J.Sm.

Distribution: Indonesia, Malaysia

Dendrobium aries J.J.Sm.

Distribution: Indonesia

Dendrobium aristiferum J.J.Sm.
Pedilonum aristiferum (J.J.Sm.) Rauschert

Distribution: Indonesia

Dendrobium armeniacum P.J.Cribb
Sayeria armeniaca (P.J.Cribb) Rauschert

Distribution: Papua New Guinea

Dendrobium asphale Rchb.f.

Distribution: Unknown

Dendrobium atavus J.J.Sm.

Distribution: Indonesia

Dendrobium atjehense J.J.Sm.

Distribution: Indonesia

Dendrobium atroviolaceum Rolfe
Latourorchis atroviolaceum (Rolfe) Brieger
Sayeria atroviolacea (Rolfe) Rauschert

Distribution: Papua New Guinea

Dendrobium aurantiroseum P.Royen ex T.M.Reeve

Distribution: Indonesia, Papua New Guinea

Dendrobium auriculatum Ames & Quisumb.

Distribution: Philippines (the)

Dendrobium babiense J.J.Sm.
  Aporum babiense (J.J.Sm.) Rauschert

Distribution: Indonesia, Malaysia

Dendrobium baeuerlenii F. Muell. & Kraenzl.
  Pedilonum baeuerlenii (F. Muell. & Kraenzl.) Rauschert

Distribution: Papua New Guinea

Dendrobium barbatulum Lindl.
  Callista barbatula (Lindl.) Kuntze

Distribution: India

Dendrobium basilanense Ames

Distribution: Philippines (the)

Dendrobium bellatulum Rolfe

Distribution: China, India, Lao People's Democratic Republic, Myanmar, Thailand, Viet Nam

Dendrobium bensoniae Rchb.f.
  Callista bensoniae (Rchb.f.) Kuntze

Distribution: India, Myanmar, Thailand

Dendrobium bicameratum Lindl.
  Callista bicamerata (Lindl.) Kuntze
  Callista bolboflora (Falc. ex Hook.f.) Kuntze
  Callista breviflora (Lindl.) Kuntze
  Dendrobium bolboflorum Falc. ex Hook.f.
  Dendrobium breviflorum Lindl.

Distribution: Bhutan, India, Myanmar, Nepal, Thailand

Dendrobium bicaudatum Reinw. ex Lindl.
  Callista bicaudata (Reinw. ex Lindl.)
  Dendrobium antelope Rchb.f.
  Dendrobium burbidgei Rchb.f.

Part II: Dendrobium

Dendrobium demmenii J.J.Sm.
Dendrobium minax Rchb.f.
Dendrobium rumphianum Teijsm. & Binn.

Distribution: Indonesia

Dendrobium bicornutum Schltr.
Aporum bicornutum (Schltr.) Rauschert

Distribution: Indonesia

Dendrobium bifalce Lindl.
Bulbophyllum oncidiochilum Kraenzl.
Callista bifalcis (Lindl.) Kuntze
Dendrobium breviracemosum F.M.Bailey
Dendrobium chloropterum Rchb.f. & S.Moore
Dendrobium chloropterum var. *striatum* J.J.Sm.
Doritis bifalcis (Lindl.) Rchb.f.
Sayeria bifalcis (Lindl.) Rauschert

Distribution: Australia, Indonesia, Papua New Guinea, Solomon Islands

Dendrobium bigibbum Lindl.
Callista biggiba (Lindl.) Kuntze

Distribution: Australia, Indonesia, Papua New Guinea

Dendrobium bilobulatum Seidenf.

Distribution: Thailand, Viet Nam

Dendrobium biloculare J.J.Sm.
Sayeria bilocularis (J.J.Sm.) Rauschert

Distribution: Indonesia

Dendrobium blumei Lindl.
Aporum blumei (Lindl.) Rauschert
Callista boothii (Teijsm. & Binn.) Kuntze
Callista fimbriata (Blume) Kuntze
Dendrobium boothii Teijsm. & Binn.
Onychium fimbriatum Blume

Distribution: Indonesia, Malaysia, Philippines (the), Thailand

Dendrobium bostrychodes Rchb.f.

Distribution: Brunei Darassalam, Malaysia

Dendrobium boumaniae J.J.Sm.

Distribution: Indonesia

Dendrobium bowmanii Benth.
Aporum chalandei (Finet) Rauschert
Dendrobium chalandei (Finet) Kraenzl.
Dendrobium striolatum var. *chalandei* Finet
Dockrilla bowmanii (Benth.) M.A.Clem. & D.L. Jones

Distribution: Australia, New Caledonia

Dendrobium brachycalyptra Schltr.
Pedilonum brachycalyptra (Schltr.) Rauschert

Distribution: Papua New Guinea

Dendrobium brachycentrum Ridl.
Pedilonum brachycentrum (Ridl.) Rauschert

Distribution: Indonesia

Dendrobium brachypus (Endl.) Rchb.f.
Callista brachypus (Rchb.f.) Kuntze
Thelychiton brachypus Endl.

Distribution: Australia

Dendrobium bracteosum Rchb.f.
Dendrobium bracteosum var. *album* Sander
Dendrobium bracteosum var. *roseum* Sander
Dendrobium chrysolabium Rolfe
Dendrobium dixsonii F.M.Bailey
Dendrobium dixsonii var. *eborinum* F.M.Bailey
Dendrobium eitapense Schltr.
Dendrobium leucochysum Schltr.
Dendrobium novae-hiberniae Kraenzl.
Dendrobium quadrialatum J.J.Sm.
Dendrobium trisaccatum Kraenzl.
Pedilonum bracteosum (Rchb.f.) Rauschert
Pedilonum eitapense (Schltr.) Rauschert
Pedilonum leucochysum (Schltr.) Rauschert
Pedilonum quadrialatum (J.J.Sm.) Rauschert

Distribution: Indonesia, Papua New Guinea

Dendrobium brassii T.M.Reeve & P.Woods

Distribution: Papua New Guinea

103

Dendrobium brevicaule Rolfe
Dendrobium cyatheicola Van Royen
Pedilonum brevicaule (Rolfe) Rauschert

Distribution: Indonesia, Papua New Guinea

Dendrobium brevicaule subsp. **brevicaule**

Distribution: Indonesia

Dendrobium brevicaule subsp. **calcarium** (J.J.Sm.) T.M.Reeve & P.Woods
Dendrobium aurantivinosum Van Royen
Dendrobium calcarium J.J.Sm.
Dendrobium montistellare Van Royen
Dendrobium quinquecristatum Van Royen
Pedilonum calcarium (J.J.Sm.) Rauschert

Distribution: Indonesia, Papua New Guinea

Dendrobium brevicaule subsp. **pentagonum** (Kraenzl.) T.M.Reeve
Dendrobium pentagonum Kraenzl.
Dendrobium saruwagedicum Schltr.
Dendrobium teligerum Van Royen
Dendrobium zaranense Van Royen
Pedilonum pentagonum (Kraenzl.) Rauschert
Pedilonum saruwagedicum (Schltr.) Rauschert

Distribution: Papua New Guinea

Dendrobium brevilabium Schltr.
Pedilonum brevilabium (Schltr.) Rauschert

Distribution: Papua New Guinea

Dendrobium brevimentum Seidenf.

Distribution: Thailand

Dendrobium brymerianum Rchb.f.
Callista brymeriana (Rchb.f.) Kuntze

Distribution: China, India, Lao People's Democratic Republic (the), Myanmar, Thailand, Viet Nam

Dendrobium brymerianum var. **brymerianum**

Distribution: China, India, Lao People's Democratic Republic, Myanmar, Thailand, Viet Nam

Dendrobium brymerianum var. **histrionicum** Rchb.f.

Distribution: Myanmar, Thailand

Dendrobium buffumii A.D. Hawkes
Dendrobium roseoflavidum Schltr.
Pedilonum buffumii (A.D. Hawkes) Rauschert

Distribution: Papua New Guinea

Dendrobium bullenianum Rchb.f.
Dendrobium chrysocephalum Kraenzl.
Dendrobium erythroxanthum Rchb.f.
Dendrobium reinwardtii auct. non Lindl.
Dendrobium salaccense auct. non Lindl.
Dendrobium topaziacum Ames

Distribution: Philippines (the)

Dendrobium cacatua M.A.Clem. & D.L.Jones
Dendrobium tetragonum var. *cacatua* (M.A.Clem. & D.L.Jones) Mohr.
Dendrobium tetragonum var. *giganteum* Leaney

Distribution: Australia

Dendrobium calamiforme Lodd.
Dendrobium baseyanum St.Cloud
Dendrobium teretifolium var. *album* C.T.White
Dendrobium teretifolium var. *fasciculatum* Rupp
Dockrillia baseyana (St.Cloud) Rauschert
Dockrillia calamiformis (Lodd.) M.A.Clem. & D.L.Jones

Distribution: Australia

Dendrobium calcaratum A.Rich.
Dendrobium achillis Rchb.f.
Dendrobium separatum Ames
Dendrobium triviale Kraenzl.
Pedilonum triviale (Kraenzl.) Rauschert
Pedilonum separatum Ames

Distribution: Papua New Guinea, Samoa, Solomon Islands, Tonga, Vanuatu

Dendrobium calcaratum subsp. **calcaratum**

Distribution: Samoa, Solomon Islands, Tonga, Vanuatu

105

Part II: Dendrobium

Dendrobium calcaratum subsp. **papillatum** Dauncey

Distribution: Papua New Guinea

Dendrobium calcariferum Carr

Distribution: Malaysia

Dendrobium calceolum Roxb.
Aporum calceolum (Roxb.) Rauschert
Aporum roxburghii Griff.
Callista calceolum (Roxb.) Kuntze
Dendrobium roxburghii Lindl.

Distribution: Indonesia

Dendrobium calicopis Ridl.
Dendrobium derryi auct. non Ridl.
Dendrobium sanguinolentum auct. non Lindl.
Pedilonum calicopis (Ridl.) Rauschert
Pedilonum derryi (Ridl.) Rauschert

Distribution: Malaysia, Thailand

Dendrobium caliculimentum R.S.Rogers
Dendrobium altomontanum Gilli
Dendrobium lompobatangense J.J.Sm.
Dendrobium rhododiodes P.Royen
Pedilonum caliculimentum (R.S.Rogers) Rauschert
Pedilonum lompobatangense (J.J.Sm.) Rauschert

Distribution: Indonesia, Papua New Guinea, Solomon Islands

Dendrobium calophyllum Rchb.f.
Callista calophylla (Rchb.f.) Kuntze
Dendrobium rimannii Rchb.f.

Distribution: Indonesia

Dendrobium calyptratum J.J.Sm.
Pedilonum calyptratum (J.J.Sm.) Rauschert

Distribution: Indonesia

Dendrobium canaliculatum R.Br.
Callista canaliculata (R.Br.) Kuntze
Dendrobium canaliculatum var. *nigrescens* Nicholls

Distribution: Australia, Papua New Guinea

106

Dendrobium canaliculatum var. canaliculatum

Distribution: Australia, Papua New Guinea

Dendrobium canaliculatum var. pallidum Dockrill

Distribution: Australia

Dendrobium capillipes Rchb.f.
Callista acrobatica (Rchb.f.) Kuntze
Callista capillipes (Rchb.f.) Kuntze
Dendrobium braianense Gagnep.

Distribution: China, India, Lao People's Democratic Republic, Myanmar, Nepal, Thailand, Viet Nam

Dendrobium capitisyork M.A.Clem. & D.L.Jones

Distribution: Australia

Dendrobium capituliflorum Rolfe
Angraecum purpureum Rumph.
Dendrobium capituliflorum var. *viride* J.J.Sm.
Dendrobium confusum J.J.Sm.
Dendrobium ophioglossum auct. non Rchb.f.
Pedilonum capituliflorum (Rolfe) Brieger

Distribution: Indonesia, Papua New Guinea, Solomon Islands, Vanuatu

Dendrobium capra J.J.Sm.

Distribution: Indonesia

Dendrobium cariniferum Rchb.f.
Callista carinifera (Rchb.f.) Kuntze

Distribution: China, India, Lao People's Democratic Republic, Myanmar, Thailand, Viet Nam

Dendrobium carrii Rupp & C.T.White
Katherinea carrii (Rupp & C.T.White) Brieger

Distribution: Australia

Dendrobium carronii Lavarack & P.J.Cribb

Distribution: Australia, Papua New Guinea

107

Part II: Dendrobium

Dendrobium caryicola Guill.

Distribution: Viet Nam

Dendrobium casuarinae Schltr.
Dockrillia casuarinae (Schltr.) M.A.Clem. & D.L. Jones

Distribution: New Caledonia

Dendrobium catillare Rchb.f.
Callista glossotis (Rchb.f.) Kuntze
Dendrobium glomeriflorum Kraenzl.
Dendrobium glossotis Rchb.f.
Dendrobium purpureum auct. non Roxb.
Dendrobium secundum auct. non (Blume) Lindl.
Dendrobium sertatum Rolfe
Pedilonum glossotis (Rchb.f.) Rauschert

Distribution: Fiji

Dendrobium ceraula Rchb.f.
Dendrobium gonzalesii Quisumb.

Distribution: Philippines (the)

Dendrobium chameleon Ames
Dendrobium longicalcaratum Hayata
Dendrobium randaiense Hayata
Pedilonum lonicalcaratum (Hayata) Rauschert

Distribution: China (Taiwan), Philippines (the)

Dendrobium chanjiangense S.J. Cheng &C.Z. Tang

Distribution: China

Dendrobium chlorostylum Gagnep.

Distribution: Viet Nam

Dendrobium chordiforme Kraenzl.
Dockrillia chordiformis (Kraenzl.) Rauschert

Distribution: Papua New Guinea

Dendrobium christyanum Rchb.f.

Dendrobium margaritaceum Finet

Distribution: Thailand, Viet Nam

Dendrobium chrysanthum Wall.
Callista chrysantha (Lindl.) Kuntze
Dendrobium paxtonii Lindl.

Distribution: Bhutan, China, India, Lao People's Democratic Republic, Myanmar, Nepal, Thailand, Viet Nam

Dendrobium chryseum Rolfe
Aporum rivesii (Gagnep.) Rauschert
Callista aurantiaca (Rchb.f.) Kuntze
Callista clavata (Lindl.) Kuntze
Dendrobium aurantiacum Rchb.f.
Dendrobium clavatum Lindl.
Dendrobium denneanum Kerr
Dendrobium flaviflorum Hayata
Dendrobium rivesii Gagnep.
Dendrobium rolfei Hawkes & Heller
Dendrobium tibeticum Schlechter

Distribution: Bhutan, China, China (Taiwan), India, Lao People's Democratic Republic, Myanmar, Nepal, Thailand, Viet Nam

Dendrobium chrysocrepis Parish & Rchb.f.
Callista chyrsocrepis (Par. & Rchb.f.) Kuntze

Distribution: India, Myanmar

Dendrobium chrysotoxum Lindl.
Callista chrysotoxa (Lindl.) Kuntze
Callista chrysotoxum var. *delacourii* Gagnep.
Callista suavissima (Rchb.f.) Kuntze
Dendrobium chrysotoxum var. *suavissimum* (Rchb.f.) Veitch
Dendrobium suavissimum Rchb.f.

Distribution: Bhutan, China, India, Lao People's Democratic Republic, Myanmar, Nepal, Thailand, Viet Nam

Dendrobium ciliatilabellum Seidenf.

Distribution: Thailand

Dendrobium cinereum J.J.Sm.
Dendrobium groeneveldtii J.J.Sm.

Distribution: Indonesia, Malaysia

109

Part II: Dendrobium

Dendrobium cinnabarinum Rchb.f.
Aporum cinnabarinum (Rchb.f.) Rauschert

Distribution: Brunei Darussalam, Malaysia

Dendrobium cinnabarinum var. **angustitepalum** Carr
Dendrobium holttumianum A.D.Hawkes & A.H.Heller
Dendrobium sanguineum Rolfe

Distribution: Malaysia

Dendrobium cinnabarinum var. **cinnabarinum**

Distribution: Brunei Darussalam, Malaysia

Dendrobium cinnabarinum var. **lamelliferum** Carr

Distribution: Malaysia

Dendrobium clavator Ridl.
Aporum clavator (Ridl.) Rauschert
Dendrobium gracile auct. non Lindl.
Dendrobium mellitum Ridl.

Distribution: Malaysia, Thailand

Dendrobium cochleatum J.J.Sm.
Pedilonum cochleatum (J.J.Sm.) Rauschert

Distribution: Papua New Guinea

Dendrobium cochliodes Schltr.
Dendrobium ruidilobum J.J.Sm.

Distribution: Indonesia, Papua New Guinea

Dendrobium codonosepalum J.J.Sm.

Distribution: Indonesia, Papua New Guinea

Dendrobium compactum Rolfe ex W.Hackett

Distribution: China, Myanmar, Thailand

Dendrobium compressimentum J.J.Sm.
Aporum compressimentum (J.J.Sm.) Rauschert

Distribution: Indonesia, Malaysia

Dendrobium comptonii Rendle
Dendrobium drake-castilloi Kraenzl.
Dendrobium macropus subsp. *howeanum* (Maiden) P.S.Green
Tropilis comptonii (Rendle) Rauschert
Tropilis drake-castilloi (Kraenzl.) Rauschert

Distribution: Australia, New Caledonia

Dendrobium conanthum Schltr.
Dendrobium busuangense Ames
Dendrobium kajewskii Ames

Distribution: Indonesia, Papua New Guinea, Solomon Islands, Vanuatu

Dendrobium confinale Kerr

Distribution: Thailand

Dendrobium conicum J.J.Sm.
Pedilonum conicum (J.J.Sm.) Rauschert

Distribution: Indonesia

Dendrobium constrictum J.J.Sm.
Dendrobium mimiense Schltr.
Pedilonum mimiense (Schltr.) Brieger

Distribution: Indonesia, Papua New Guinea

Dendrobium convexipes J.J.Sm.

Distribution: Indonesia

Dendrobium convolutum Rolfe
Sayeria convoluta (Rolfe) Rauschert

Distribution: Papua New Guinea

Dendrobium corallorhizon J.J.Sm.

Distribution: Indonesia

Dendrobium crabro Ridl.

Part II: Dendrobium

Distribution: Malaysia

Dendrobium crenatifolium J.J.Sm.
Pedilonum crenatifolium (J.J.Sm.) Rauschert

Distribution: Indonesia

Dendrobium crepidatum Lindl. & Paxton
Callista crepidata (Lindl. & Paxton) Kuntze
Callista lawana (Lindl.) Kuntze
Dendrobium actinomorphum Blatter & Hallb.
Dendrobium crepidatum var. *avista* Gammie
Dendrobium lawanum Lindl.
Dendrobium roseum Dalzell

Distribution: Bhutan, Cambodia, China, India, Lao People's Democratic Republic, Myanmar, Nepal, Thailand, Viet Nam

Dendrobium cretaceum Lindl.
Callista cretacea (Lindl.) Kuntze
Dendrobium polyanthum Lindl.

Distribution: China, India, Lao People's Democratic Republic, Myanmar, Nepal, Thailand, Viet Nam

Dendrobium crispilinguum P.J.Cribb

Distribution: Papua New Guinea

Dendrobium crocatum Hook.f.
Callista crocata (Hook.f.) Kuntze
Dendrobium pyropum Ridl.
Pedilonum crocatum (Hook.f.) Brieg.
Pedilonum pyropum (Ridl.) Rauschert

Distribution: Malaysia, Thailand

Dendrobium croceocentrum J.J.Sm.

Distribution: Indonesia

Dendrobium crucilabre J.J.Sm.
Aporum crucilabre (J.J.Sm.) Rauschert

Distribution: Indonesia

Dendrobium cruentum Rchb.f.
Callista cruenta (Rchb.f.) Kuntze

Distribution: Malaysia, Myanmar, Thailand

Dendrobium crumenatum Sw.
Angraecum crumenatum Rumph.
Aporum crumenatum (Sw.) Brieger
Aporum kwashotense (Hayata) Rauschert
Callista crumenata (Sw.) Kuntze
Ceraia simplicissima Lour.
Dendrobium caninum sensu Merr.
Dendrobium ceraia Lindl.
Dendrobium crumenatum var. *parviflorum* Ames & C.Schweinf.
Dendrobium kwashotense Hayata
Dendrobium schmidtianum Kraenzl.
Dendrobium simplicissimum (Lour.) Kraenzl.
Epidendrum spatulatum auct. non Burm.f.
Onychium crumenatum (Sw.) Blume

Distribution: Australia, Bangladesh, Brunei Darussalam, Cambodia, China, China (Taiwan), India, Indonesia, Lao People's Democratic Republic, Malaysia, Myanmar, Philippines (the), Seychelles, Singapore, Sri Lanka, Thailand, Viet Nam

Dendrobium cruttwellii T.M.Reeve
Dendrobium sayeria Schltr.
Sayeria paradoxa Kraenzl.

Distribution: Indonesia, Papua New Guinea

Dendrobium crystallinum Rchb.f.
Callista crystallina (Rchb.f.) Kuntze

Distribution: Cambodia, China, Lao People's Democratic Republic, Myanmar, Thailand, Viet Nam

Dendrobium cuculliferum J.J.Sm.
Pedilonum cuculliferum (J.J.Sm.) Rauschert

Distribution: Indonesia

Dendrobium cucumerinum Macleay ex Lindl.
Callista cucumerinum (Macleay ex Lindl.) Kuntze
Dockrillia cucumerina (Macleay ex Lindl.) Brieger

Distribution: Australia

Dendrobium cumulatum Lindl.
Callista cumulata (Lindl.) Kuntze
Dendrobium eoum Ridl.
Dendrobium intricatum auct. non Gagnep.
Dendrobium sanguinolentum auct. non Lindl.

Part II: Dendrobium

Pedilonum eoum (Ridl.) Rauschert

Distribution: Bhutan, Cambodia, India, Indonesia, Lao People's Democratic Republic, Malaysia, Myanmar, Nepal, Philippines, Thailand, Viet Nam

Dendrobium cunninghamii Lindl.
Callista cunninghamii (Lindl.) Kuntze

Distribution: New Zealand

Dendrobium curvicaule (F.M.Bailey) M.A.Clem. & D.L.Jones
Dendrobium speciosum var. *curvicaule* F.M.Bailey

Distribution: Australia

Dendrobium curviflorum Rolfe

Distribution: India, Myanmar, Thailand

Dendrobium curvimentum J.J.Sm.
Sayeria curvimenta (J.J.Sm.) Rauschert

Distribution: Indonesia

Dendrobium cuspidatum Lindl.
Callista cuspidata (Lindl.) Kuntze
Callista sarcantha (Lindl.) Kuntze
Dendrobium sarcanthum Lindl.

Distribution: Myanmar, Thailand

Dendrobium cuthbertsonii F.Muell.
Dendrobium agathodaemonis J.J.Sm.
Dendrobium asperifolium J.J.Sm.
Dendrobium atromarginatum J.J.Sm.
Dendrobium carstensziense J.J.Sm.
Dendrobium coccinellum Ridl.
Dendrobium euphues Ridl.
Dendrobium fulgidum Ridl.
Dendrobium fulgidum var. *purpureum* Ridl.
Dendrobium laetum Schltr.
Dendrobium lichenicola J.J.Sm.
Dendrobium sophronites Schltr.
Dendrobium trachyphyllum Schltr.
Pedilonum asperifolium (J.J.Sm.) Brieger
Pedilonum coccinellum (Ridl.) Rauschert
Pedilonum cuthbertsonii (F.Muell.) Brieger
Pedilonum euphues (Ridl.) Rauschert
Pedilonum sophronites (Schltr.) Rauschert
Pedilonum trachyphyllum (Schltr.) Brieger

Distribution: Indonesia, Papua New Guinea

Dendrobium cyanocentrum Schltr.
Dendrobium flavispiculum J.J.Sm.
Dendrobium lapeyrouseiodes Schltr.
Pedilonum cyanocentrum (Schltr.) Rauschert
Pedilonum flavispiculum (J.J.Sm.) Rauschert
Pedilonum lapeyrouseoides (Schltr.) Rauschert.

Distribution: Indonesia, Papua New Guinea

Dendrobium cylindricum J.J.Sm.
Pedilonum cylindricum (J.J.Sm.) Rauschert

Distribution: Indonesia

Dendrobium cymboglossum J.J.Wood & A.Lamb

Distribution: Malaysia

Dendrobium cymbulipes J.J.Sm.
Aporum cymbulipes (J.J.Sm.) Rauschert

Distribution: Indonesia, Malaysia

Dendrobium dalatense Gagnep.
Aporum dalatense (Gagnep.) Rauschert

Distribution: Viet Nam

Dendrobium dantaniense Guill.
Dendrobium alterum Seidenf.

Distribution: Thailand, Viet Nam

Dendrobium daoense Gagnep.

Distribution: Viet Nam

Dendrobium darjeelingensis Pradhan

Distribution: India

Dendrobium dearei Rchb.f.
Callista dearei (Rchb.f.) Kuntze

Distribution: Philippines (the)

Dendrobium dekockii J.J.Sm.
Dendrobium cedricola P. Van Royen
Dendrobium chrysornis Ridl.
Dendrobium erythocarpum J.J.Sm.
Dendrobium gaudens P. Van Royen
Dendrobium kerewense P. Van Royen
Dendrobium montigenum Ridl.
Pedilonum chrysornis (Ridl.) Rauschert
Pedilonum dekockii (J.J.Sm.) Rauschert
Pedilonum montigenum (Ridl.) Rauschert

Distribution: Indonesia, Papua New Guinea

Dendrobium delacourii Guill.
Dendrobium ciliatum auct. non Pers.
Dendrobium ciliatum var. *breve* Rchb.f.
Dendrobium rupicola var. *breve* (Rchb.f.) Atwood

Distribution: Cambodia, Lao People's Democratic Republic, Myanmar, Thailand, Viet Nam

Dendrobium delicatulum Kraenzl.
Dendrobium minutum Schltr.
Dendrobium nanarauticola Fukuy.
Pedilonum delicatulum (Kraenzl.) Rauschert
Pedilonum minutum (Schltr.) Rauschert

Distribution: Fiji, Indonesia, Micronesia (Federated States of), Papua New Guinea, Solomon Islands, Vanuatu

Dendrobium delicatulum subsp. **delicatulum**

Distribution: Fiji, Indonesia, Micronesia (Federated States of), Papua New Guinea, Solomon Islands, Vanuatu

Dendrobium delicatulum subsp. **huliorum** T.M.Reeve & P.Woods

Distribution: Papua New Guinea

Dendrobium delicatulum subsp. **parvulum** (Rolfe) T.M.Reeve & P.Woods
Dendrobium parvulum Rolfe
Katherinea parvula (Rolfe) A.D.Hawkes
Sarcopodium parvulum (Rolfe) Kraenzl.

Distribution: Indonesia

Dendrobium × delicatum (F.M.Bailey) F.M.Bailey
Dendrobium speciosum var. *album* B.S.Williams
Dendrobium speciosum var. *delicatum* F.M.Bailey

Tropilis × *delicata* (F.M.Bailey) Butzin

Distribution: Australia

Dendrobium deltatum Seidenf.
Dendrobium trinervium auct. non. Ridl.

Distribution: Lao People's Democratic Republic, Thailand

Dendrobium dendrocolloides J.J.Sm.
Dendrobium incurvilabium Schltr.
Sayeria incurvilabia (Schltr.) Rauschert

Distribution: Indonesia, Papua New Guinea

Dendrobium densiflorum Wall. ex Lindl.
Callista densiflora (Lindl.) Kuntze
Dendrobium clavatum Roxb.
Dendrobium griffithianum var. *guibertii* (Carriere) Veitch
Dendrobium guibertii Carriere

Distribution: Bhutan, China, India, Lao People's Democratic Republic, Myanmar, Nepal, Thailand

Dendrobium dentatum Seidenf.

Distribution: Viet Nam

Dendrobium denudans D.Don
Callista denudans (D.Don) Kuntze

Distribution: India, Nepal

Dendrobium devonianum Paxton
Callista devoniana (Paxt.) Kuntze
Dendrobium pictum Griff. ex Lindl.
Dendrobium pulchellum var. *devonianum* (Paxt.) Rchb.f.

Distribution: Bhutan, China, India, Myanmar, Thailand, Viet Nam

Dendrobium devosianum J.J.Sm.

Distribution: Indonesia

Dendrobium diceras Schltr.
Sayeria diceras (Schltr.) Rauschert

Distribution: Papua New Guinea

Dendrobium dichaeoides Schltr.
Pedilonum dichaeoides (Schltr.) Rauschert

Distribution: Papua New Guinea

Dendrobium dichroma Schltr.
Pedilonum dichroma (Schltr.) Rauschert

Distribution: Papua New Guinea

Dendrobium dickasonii L.O.Williams
Callista arachnites (Rchb.f.) Kuntze
Dendrobium arachnites Rchb.f.
Dendrobium seidenfadenii Senghas & Bockemühl

Distribution: India, Myanmar, Thailand

Dendrobium dillonianum A.D.Hawkes & A.H.Heller
Dendrobium sacculiferum J.J.Sm.

Distribution: Indonesia, Papua New Guinea

Dendrobium discolor Lindl
Callista undulata (R.Br.) Kuntze
Dendrobium arachnanthe Kraenzl.
Dendrobium elobatum Rupp
Dendrobium undulatum R.Br

Distribution: Australia, Papua New Guinea

Dendrobium discolor var. **broomfieldii** (Fitzg.) M.A.Clem. & D.L.Jones
Dendrobium broomfieldii (Fitzg.) Fitzg.
Dendrobium undulatum var. *broomfieldii* Fitzg.

Distribution: Australia

Dendrobium discolor var. **discolor**

Distribution: Australia, Papua New Guinea

Dendrobium discolor var. **fimbrilabium** (Rchb.f.) Dockrill
Dendrobium undulatum var. *fimbrilabium* Rchb.f.

Distribution: Australia

Dendrobium discolor var. **fuscum** (Fitzg.) Dockrill

Distribution: Australia

Dendrobium dixanthum Rchb.f.
Callista dixantha (Rchb.f.) Kuntze

Distribution: China, Lao People's Democratic Republic, Myanmar, Thailand

Dendrobium dixonianum Rolfe ex Downie
Dendrobium eriiflorum auct. non Griff.

Distribution: Thailand

Dendrobium dolichophyllum D.L.Jones & M.A.Clem.
Dendrobium teretifolium var. *aureum* F.M.Bailey
Dockrillia dolichophylla (M.A.Clem. & D.L.Jones) M.A.Clem. & D.L.Jones

Distribution: Australia

Dendrobium draconis Rchb.f.
Callista draconis (Rchb.f.) Kuntze
Dendrobium andersonii Scott.
Dendrobium eburneum Rchb.f. ex Bateman

Distribution: Cambodia, India, Lao People's Democratic Republic, Myanmar, Thailand, Viet Nam

Dendrobium eboracense Kraenzl.
Aporum eboracense (Kraenzl.) Rauschert
Dendrobium macfarlanei F.Muell.

Distribution: Papua New Guinea

Dendrobium endertii J.J.Sm.

Distribution: Indonesia

Dendrobium engae T.M.Reeve

Distribution: Papua New Guinea

Dendrobium eriiflorum Griff.
Callista eriaeflora (Griff.) Kuntze

Distribution: Bhutan, India, Indonesia, Malaysia, Myanmar, Nepal, Thailand

Dendrobium erostelle Seidenf.

119

Part II: Dendrobium

Distribution: Thailand

Dendrobium erosum (Blume) Lindl.
Callista erosa (Blume) Kuntze
Dendrobium aegle Ridl.
Dendrobium aemulans Schltr.
Dendrobium inopinatum J.J.Sm.
Pedilonum aegle (Ridl.) Rauschert
Pedilonum aemulans (Schltr.) Rauschert
Pedilonum erosum Blume
Pedilonum inopinatum (J.J.Sm.) Rauschert

Distribution: Indonesia, Malaysia, Solomon Islands, Thailand, Vanuatu

Dendrobium erythropogon Rchb.f.

Distribution: Malaysia

Dendrobium escritorii Ames

Distribution: Philippines (the)

Dendrobium eserre Seidenf.

Distribution: Thailand

Dendrobium eumelinum Schltr.
Pedilonum eumelinum (Schltr.) Rauschert

Distribution: Papua New Guinea

Dendrobium euryanthum Schltr.
Sayeria euryantha (Schltr.) Rauschert

Distribution: Papua New Guinea

Dendrobium exile Schltr.
Aporum heterocaulon (Guill.) Rauschert
Dendrobium heterocaulon Guill.
Dendrobium tetraedre auct. non Lindl.

Distribution: China, Thailand, Viet Nam

Dendrobium eximium Schltr.
Dendrobium bellum J.J.Sm.
Dendrobium wollastonii Ridl.
Sayeria eximia (Schltr.) Rauschert

Distribution: Indonesia, Papua New Guinea

120

Dendrobium fairfaxii F.Muell. & Fitzg.
Callista fairfaxii (F.Muell. & Fitzg.) Kuntze
Dendrobium teretifolium var. *fairfaxii* (F.Muell. & Fitzg.) F.M.Bailey
Dockrillia fairfaxii (F.Muell. & Fitzg.) Rauschert

Distribution: Australia

Dendrobium falconeri Hook.
Callista falconeri (Hook.) Kuntze
Dendrobium erythroglossum Hayata

Distribution: Bhutan, China, China (Taiwan), India, Myanmar, Thailand

Dendrobium falcorostrum Fitzg.
Callista falcorostris (Fitzg.) Kuntze
Tropilis falcorostra (Fitzg.) Rauschert
Tropilis falcorostra (Fitzg.) Butzin

Distribution: Australia

Dendrobium farmeri Paxton
Callista farmeri (Paxt.) Kuntze
Dendrobium densiflorum var. *alboluteum* Hook.f.

Distribution: Bhutan, India, Lao People's Democratic Republic, Malaysia, Myanmar, Nepal, Thailand, Viet Nam

Dendrobium faulhaberianum Schltr.
Dendrobium aduncum auct. non Lindl.
Dendrobium aduncum var. *faulhaberianum* (Schltr.) Tang. & Wang.
Dendrobium hercoglossum auct. non Rchb.f.
Dendrobium oxyanthum Gagnep.

Distribution: China, Lao People's Democratic Republic, Viet Nam

Dendrobium fellowsii F.Muell.
Callista bairdiana (F.M.Bailey) Kuntze
Dendrobium bairdianum F.M.Bailey
Dendrobium giddinsii T.E.Hunt
Sayeria bairdiana (F.M.Bailey) Rauschert

Distribution: Australia

Dendrobium fesselianum M.Wolff

Distribution: Thailand

121

Part II: Dendrobium

Dendrobium filicaule Gagnep.

Distribution: Viet Nam

Dendrobium fimbriatum Hook.
Callista oculata (Hook.) Kuntze
Dendrobium fimbriatum var. *oculatum* Hook.
Dendrobium hawkesii Heller
Dendrobium paxtonii Paxton
Dendrobium vagans Gagnep.

Distribution: Bhutan, China, India, Lao People's Democratic Republic, Malaysia, Myanmar, Nepal, Thailand, Viet Nam

Dendrobium fimbrilabium J.J.Sm.
Pedilonum fimbrilabium (J.J.Sm.) Rauschert

Distribution: Indonesia

Dendrobium findlayanum Parish & Rchb.f.
Callista findlayana (Parish & Rchb.f.) Kuntze

Distribution: China, Lao People's Democratic Republic, Myanmar, Thailand

Dendrobium finisterrae Schltr.
Dendrobium finisterrae var. *polystichum* Schltr.
Dendrobium melanolasium Gilli
Sayeria finisterrae (Schltr.) Rauschert

Distribution: Indonesia, Papua New Guinea

Dendrobium flagellum Schltr.
Dockrillia flagella (Schltr.) Rauschert

Distribution: Papua New Guinea

Dendrobium flammula Schltr.
Pedilonum flammula (Schltr.) Rauschert

Distribution: Papua New Guinea

Dendrobium fleckeri Rupp & C.T.White
Tropilis fleckeri (Rupp & C.T.White) Butzin
Tropilis fleckeri (Rupp & C.T.White) Rauschert

Distribution: Australia

Dendrobium flexicaule Z.H. Tsi, S.c. Sum & L.C.Xu

Distribution: China

Dendrobium × foederatum St.Cloud
Dockrillia × foederata (St.Cloud) Rauschert

Distribution: Australia

Dendrobium foelschei F.Muell.
Callista foelschei (F.Muell.) Kuntze

Distribution: Australia

Dendrobium forbesii Ridl.
Dendrobium ashworthiae O'Brien
Dendrobium eustachyum Schltr.
Dendrobium forbesii var. *praestans* Schltr.
Latourorchis forbesii (Ridl.) Brieger
Sayeria eustachya (Schltr.) Rauschert
Sayeria forbesii (Ridl.) Rauschert

Distribution: Papua New Guinea

Dendrobium formosum Roxb. ex Lindl.
Callista formosa (Roxb. ex Lindl.) Kuntze

Distribution: Bhutan, India, Myanmar, Nepal, Thailand

Dendrobium fractum T.M.Reeve

Distribution: Papua New Guinea

Dendrobium friedericksianum Rchb.f.
Dendrobium friedericksianum var. *oculatum* Seidenf. & Smitin.

Distribution: Lao People's Democratic Republic, Thailand

Dendrobium fruticicola J.J.Sm.
Pedilonum fruticicola (J.J.Sm.) Rauschert

Distribution: Indonesia

Dendrobium fuerstenbergianum Schltr.

Distribution: Thailand

Dendrobium fulgidum Schltr.

Part II: Dendrobium

Dendrobium fulgidum var. *angustilabre* J.J.Sm.
Dendrobium xanthellum Ridl.
Pedilonum fulgidum (Schltr.) Rauschert

Distribution: Indonesia, Papua New Guinea

Dendrobium fulgidum var. **fulgidum**

Distribution: Indonesia, Papua New Guinea

Dendrobium fulgidum var. **maritimum** (J.J.Sm.) Dauncey
Dendrobium maritimum J.J.Sm.

Distribution: Indonesia

Dendrobium fulminicaule J.J.Sm.

Distribution: Indonesia

Dendrobium furcatopedicellatum Hayta

Distribution: China (Taiwan)

Dendrobium fytchianum Bateman
Callista fytchiana (Bateman) Kuntze

Distribution: Myanmar

Dendrobium garrettii Seidenf.

Distribution: Thailand

Dendrobium gemellum Lindl.
Callista biflora (Blume) Kuntze
Pedilonum biflorum Blume

Distribution: Indonesia

Dendrobium geotropum T.M. Reeve

Distribution: Papua New Guinea

Dendrobium gibsonii Lindl.
Callista gibsonii (Lindl.) Kuntze
Dendrobium fimbriatum var. *gibsonii* (Lindl.) Finet
Dendrobium fuscatum Lindl.

Distribution: Bhutan, China, India, Myanmar, Nepal, Thailand

Dendrobium glaucoviride J.J.Sm.
 Pedilonum glaucoviride (J.J.Sm.) Rauschert

Distribution: Indonesia

Dendrobium glomeratum Rolfe
 Dendrobium crepidiferum J.J.Sm.
 Pedilonum crepidiferum (J.J.Sm.) Rauschert
 Pedilonum glomeratum (Rolfe) Rauschert

Distribution: Indonesia

Dendrobium gnomus Ames
 Pedilonum gnomus (Ames) Rauschert

Distribution: Solomon Islands

Dendrobium goldfinchii F.Muell.
 Aporum goldfinchii (F.Muell.) Brieger

Distribution: Papua New Guinea, Solomon Islands, Vanuatu

Dendrobium goldschmidtianum Kraenzl.
 Dendrobium irayense Ames & Quisumb.
 Pedilonum goldschmidtianum (Kraenzl.) Rauschert

Distribution: Philippines (the)

Dendrobium gouldii Rchb.f.
 Callista gouldii (Rchb.f.) Kuntze
 Dendrobium gouldii var. *acutum* Rchb.f.
 Dendrobium imthurnii Rolfe
 Dendrobium undulatum var. *woodfordianum* Maiden
 Dendrobium woodfordianum (Maiden) Schltr.

Distribution: Papua New Guinea, Solomon Islands

Dendrobium gracile (Blume) Lindl.
 Aporum gracile (Blume) Brieger
 Callista gracilis (Blume) Kuntze
 Dendrobium gedeanum J.J.Sm.
 Onychium gracile Blume

Distribution: Indonesia, Malaysia, Thailand

Dendrobium gracilicaule F.Muell.

125

Part II: Dendrobium

Callista gracilicaulis (F.Muell.) Kuntze
Tropilis gracilicaulis (F.Muell.) Butzin

Distribution: Australia

Dendrobium × gracillimum (Rupp) Leaney
Dendrobium speciosum var. *gracillimum* Rupp
Dendrobium × gracilosum Clemesha
Tropilis × gracillimum (Rupp) Butzin
Tropilis × gracillimum (Rupp) Rauschert

Distribution: Australia

Dendrobium graminifolium Lindl.
Callista graminifolia (Wight) Kuntze
Dendrobium wightii A.D.Hawkes & A.H.Heller

Distribution: India

Dendrobium grande Hook.f.
Aporum grande (Hook.f.) Rauschert
Callista grandis (Hook.f.) Kuntze

Distribution: India, Indonesia, Malaysia, Myanmar, Thailand

Dendrobium grastidioides J.J.Sm.

Distribution: Indonesia

Dendrobium gratiosissimum Rchb.f.
Callista boxallii (Rchb.f.) Kuntze
Callista gratiosissima (Rchb.f.) Kuntze
Dendrobium boxallii Rchb.f.
Dendrobium bullerianum Bateman

Distribution: China, India, Lao People's Democratic Republic, Myanmar, Thailand, Viet Nam

Dendrobium gregulus Seidenf.
Dendrobium alpestre auct. non Royle
Dendrobium compactum auct. non Rolfe

Distribution: Myanmar, Thailand

Dendrobium griffithianum Lindl.
Callista griffthiana (Lindl.) Kuntze
Dendrobium farmeri var. *aureoflava* Hook.f.

Distribution: India, Myanmar, Nepal, Thailand

Dendrobium × grimesii C.T.White & Summerh.
Dockrillia × grimesii (C.T.White & Summerh.) Rauschert

Distribution: Australia

Dendrobium grootingsii J.J.Sm.
Aporum grootingsii (J.J.Sm.) Rauschert

Distribution: Indonesia

Dendrobium guangxiense S.J. Ceng & C.Z Tang

Distribution: China

Dendrobium gynoglottis Carr

Distribution: Malaysia

Dendrobium habbemense Van Royen
Dendrobium spathulatilabratum Van Royen

Distribution: Indonesia, Papua New Guinea

Dendrobium hainanense Rolfe
Aporum hainanense (Rolfe) Rauschert

Distribution: China, Thailand, Viet Nam

Dendrobium hallieri J.J.Sm.

Distribution: Indonesia

Dendrobium hamaticalcar J.J.Wood & Dauncey

Distribution: Malaysia

Dendrobium hamatum Rolfe

Distribution: Viet Nam

Dendrobium hamiferum P.J.Cribb

Distribution: Indonesia, Papua New Guinea

Dendrobium hancockii Rolfe

Part II: Dendrobium

Distribution: China

Dendrobium harveyanum Rchb.f.
Callista harveyana (Rchb.f.) Kuntze

Distribution: China, Myanmar, Thailand, Viet Nam

Dendrobium hasseltii (Blume) Lindl.
Callista cornuta (Hook.f.) Kuntze
Callista hasseltii (Blume) Kuntze
Callista kuhlii (Blume) Kuntze
Dendrobium brinchangense Holttum
Dendrobium cornutum Hook.f.
Dendrobium curtisii Rchb.f.
Dendrobium kuhlii (Blume) Lindl.
Pedilonum brinchangense (Holttum) Rauschert
Pedilonum cornutum (Hook.f.) Rauschert
Pedilonum hasseltii Blume
Pedilonum kuhlii Blume

Distribution: Indonesia, Malaysia

Dendrobium helix P.J.Cribb

Distribution: Papua New Guinea

Dendrobium hellwigianum Kraenzl.
Dendrobium cyananthum Williams
Dendrobium geluanum Schltr.
Dendrobium rhaphiotes Schltr.
Pedilonum geluanum (Schltr.) Rauschert
Pedilonum hellwigianum (Kraenzl.) Rauschert
Pedilonum rhaphiotes (Schltr.) Rauschert

Distribution: Indonesia, Papua New Guinea

Dendrobium hemimelanoglossum Guill.

Distribution: Viet Nam

Dendrobium hendersonii A.D.Hawkes & A.H.Heller
Aporum hendersonii (Hawkes & Heller) Rauschert
Dendrobium blumei auct. non Lindl.
Dendrobium fugax Schltr.
Dendrobium hendersonii (A.D.Hawkes & A.H.Heller) Rauschert
Dendrobium ridleyanum Kerr
Dendrobium rudolphii A.D.Hawkes & A.H.Heller

Distribution: Indonesia, Malaysia, Thailand, Viet Nam

Dendrobium henryi Schltr.

Distribution: China, Thailand

Dendrobium herbaceum Lindl.
 Callista herbacea (Lindl.) Kuntze
 Dendrobium ramosissimum Wight

Distribution: India

Dendrobium hercoglossum Rchb.f.
 Callista amabilis auct. non Lour.
 Callista annamensis Kraenzl.
 Callista hercoglossa (Rchb.f.) Kuntze
 Callista vexans (Dammer) Kraenzl.
 Dendrobium linguella auct. non Rchb.f.
 Dendrobium poilanei Guill.
 Dendrobium vexans Dammer
 Dendrobium wangii C.L.Tso

Distribution: China, India, Lao People's Democratic Republic, Malaysia, Thailand, Viet Nam

Dendrobium heterocarpum Lindl.
 Callista aurea (Lindl.) Kuntze
 Callista heterocarpa (Wall. ex Lindl.) Kuntze
 Dendrobium atractodes Ridl.
 Dendrobium aureum Lindl.
 Dendrobium hildebrandii auct non Rolfe
 Dendrobium minahassae Kraenzl.
 Dendrobium rhombeum Lindl.

Distribution: Bhutan, China, India, Indonesia, Lao People's Democratic Republic, Malaysia, Myanmar, Nepal, Philippines (the), Sri Lanka, Thailand, Viet Nam

Dendrobium heyneanum Lindl.
 Callista heyneana (Lindl.) Kuntze

Distribution: India

Dendrobium hodgkinsonii Rolfe
 Sayeria hodgkinsonii (Rolfe) Rauschert

Distribution: Papua New Guinea

Dendrobium hookerianum Lindl.
 Callista hookeriana (Lindl.) Kuntze

Distribution: China, India

Part II: Dendrobium

Dendrobium hornei S.Moore ex Baker
Callista hornei (S.Moore ex Baker) Kuntze

Distribution: Fiji

Dendrobium huoshanense C.Z.Tang & S.J.Cheng

Distribution: China

Dendrobium hymenocentrum Schltr.
Aporum hymenocentrum (Schltr.) Rauschert

Distribution: Papua New Guinea

Dendrobium hymenophyllum Lindl.
Callista hymenophylla (Lindl.) Kuntze
Pedilonum undulatum Blume

Distribution: Indonesia

Dendrobium hymenopterum Hook.f.
Callista hymenoptera (Hook.f.) Kuntze
Dendrobium hymenanthum Hook.f. non Rchb.f.
Dendrobium singalanense Kraenzl.
Pedilonum hymenopterum (Hook.f.) Rauschert

Distribution: Indonesia, Malaysia, Thailand

Dendrobium igneoniveum J.J.Sm.

Distribution: Indonesia

Dendrobium inamoenum Kraenzl.
Pedilonum inamoenum (Kraenzl.) Rauschert

Distribution: Papua New Guinea

Dendrobium incurvociliatum J.J.Sm.
Aporum incurvociliatum (J.J.Sm.) Rauschert

Distribution: Indonesia, Malaysia

Dendrobium incurvum Lindl.
Callista incurva (Lindl.) Kuntze
Dendrobium aclinia Lindl.

Distribution: Cambodia, Malaysia, Myanmar, Thailand, Viet Nam

Dendrobium indivisum (Blume) Miq.
Aporum incrassatum Blume
Aporum indivisum Blume
Callista eulophota (Lindl.) Kraenzl.
Callista incrassata (Blume) Kuntze
Callista indivisa (Blume) Kraenzl.
Dendrobium eulophotum Lindl.
Dendrobium incrassatum (Blume) Miq.

Distribution: Indonesia, Lao People's Democratic Republic, Malaysia, Myanmar, Thailand, Viet Nam

Dendrobium indivisum var. **indivisum**

Distribution: Indonesia, Lao People's Democratic Republic, Malaysia, Myanmar, Thailand, Viet Nam

Dendrobium indivisum var. **pallidum** Seidenf.

Distribution: Malaysia, Myanmar, Thailand

Dendrobium inflatum Rolfe

Distribution: Indonesia

Dendrobium informe J.J.Sm.
Sayeria informis (J.J.Sm.) Rauschert

Distribution: Indonesia

Dendrobium infractum J.J.Sm.
Pedilonum infractum (J.J.Sm.) Rauschert

Distribution: Indonesia

Dendrobium infundibulum Lindl.
Callista infundibulum (Lindl.) Kuntze
Dendrobium infundibulum var. *jamesianum* (Lindl.) Veitch
Dendrobium jamesianum Rchb.f.

Distribution: China, India, Lao People's Democratic Republic, Myanmar, Thailand

Dendrobium intricatum Gagnep.
Dendrobium hymenopterum auct. non Hook.f.
Dendrobium prostratum auct. non Ridl.

Part II: Dendrobium

Distribution: Cambodia, Thailand, Viet Nam

Dendrobium jabiense J.J.Sm.
Pedilonum jabiense (J.J.Sm.) Rauschert

Distribution: Indonesia

Dendrobium jacobsonii J.J.Sm.
Pedilonum jacobsonii (J.J.Sm.) Rauschert

Distribution: Indonesia

Dendrobium jenkinsii Wall. ex Lindl.
Dendrobium aggregatum var. *jenkinsii* (Wall. ex Lindl.) King & Pantl.
Dendrobium marseilleii Gagnep.

Distribution: Bhutan, China, India, Lao People's Democratic Republic, Myanmar, Thailand

Dendrobium johannis Rchb.f.
Callista johannis (Rchb.f.) Kuntze
Dendrobium undulatum var. *johannis* (Rchb.f.) F.M.Bailey

Distribution: Australia, Indonesia, Papua New Guinea

Dendrobium johnsoniae F.Muell.
Callista macfarlanei (Rchb.f.) Kuntze
Dendrobium macfarlanei Rchb.f. non F.Muell.
Dendrobium monodon Kraenzl.
Dendrobium niveum Rolfe
Sayeria johnsoniae (F.Muell.) Rauschert

Distribution: Indonesia, Papua New Guinea, Solomon Islands

Dendrobium jonesii Rendle
Dendrobium fusiforme (F.M.Bailey) F.M.Bailey non Thouars
Dendrobium ruppianum A.D.Hawkes
Dendrobium speciosum var. *fusiforme* F.M.Bailey
Tropilis ruppiana (A.D.Hawkes) Butzin

Distribution: Australia

Dendrobium jonesii subsp. **bancroftianium** (Rchb.f.) M.A.Clem. & D.L.Jones
Dendrobium speciosum var. *bancroftianium* Rchb.f.

Distribution: Australia

Dendrobium jonesii subsp. **blackburnii** (Nicholls) M.A.Clem. & D.L.Jones
Dendrobium ruppianum var. *blackburnii* (Nicholls) Dockrill

Distribution: Australia

Dendrobium jonesii subsp. jonesii

Distribution: Australia

Dendrobium junceum Lindl.
Aporum junceum (Lindl.) Rauschert
Callista juncea (Lindl.) Kuntze

Distribution: Malaysia, Philippines (the)

Dendrobium juncoideum P. Royen

Distribution: Papua New Guinea

Dendrobium kanburiense Seidenf.

Distribution: Thailand

Dendrobium kauldorumii T.M.Reeve

Distribution: Indonesia, Papua New Guinea

Dendrobium keithii Ridl.
Dendrobium anceps auct. non Sw.
Dendrobium grande auct. non Hook.f.

Distribution: Thailand

Dendrobium × kestevenii Rupp
Dendrobium kestevenii var. *coloratum* Rupp
Dendrobium kingianum var. *kestevenii* F.M.Bailey

Distribution: Australia

Dendrobium keytsianum J.J.Sm.
Pedilonum keytsianum (J.J.Sm.) Rauschert

Distribution: Indonesia

Dendrobium kiauense Ames & C.Schweinf.
Dendrobium rajanum J.J.Sm.

Distribution: Indonesia, Malaysia

Dendrobium kingianum Bidwill ex Lindl.
Callista kingiana (Bidwill ex Lindl.) Kuntze
Dendrobium kingianum var. *aldersoniae* F.M.Bailey
Dendrobium kingianum var. *pallidum* F.M.Bailey
Dendrobium kingianum var. *pulcherrimum* Rupp
Dendrobium kingianum var. *silcockii* F.M.Bailey
Tropilis kingiana (Bidwill ex Lindl.) Butzin
Tropilis kingiana (Bidwill ex Lindl.) Rauschert

Distribution: Australia

Dendrobium klabatense Schltr.
Pedilonum klabatense (Schltr.) Rauschert

Distribution: Indonesia

Dendrobium korthalsii J.J.Sm.
Aproum korthalsii (J.J.Sm.) Rauschert

Distribution: Indonesia

Dendrobium kraemeri Schltr.
Dendrobium kraemeri var. *pseudokraemeri* (Fukuy.) Lane
Dendrobium pseudo-kraemeri Fukuy.

Distribution: Micronesia (Federated States of)

Dendrobium kratense Kerr

Distribution: Thailand

Dendrobium kruiense J.J.Sm.

Distribution: Indonesia

Dendrobium laevifolium Stapf
Dendrobium occultum Ames
Pedilonum occultum (Ames) Rauschert

Distribution: Papua New Guinea, Solomon Islands, Vanuatu

Dendrobium lamellatum (Blume) Lindl.
Callista lamellata (Blume) Kuntze
Dendrobium compressum Lindl.
Onychium lamellatum Blume
Pedilonum lamellatum (Blume) Brieger

Distribution: Indonesia, Lao People's Democratic Republic, Malaysia, Myanmar, Thailand

Dendrobium lamelluliferum J.J.Sm.

Distribution: Indonesia, Malaysia

Dendrobium lamii J.J.Sm.

Distribution: Indonesia

Dendrobium lampongense J.J.Sm.

Distribution: Indonesia, Malaysia

Dendrobium lancifolium A. Rich.
Callista lancifolia (A. Rich.) Kuntze
Dendrobium lilacinum Teijsm. & Binn.
Dendrobium vulcanicum Schltr.

Distribution: Indonesia

Dendrobium lancilabium J.J.Sm.
Pedilonum lancilabium (J.J.Sm.) Rauschert

Distribution: Indonesia

Dendrobium lancilobum J.J.Wood

Distribution: Indonesia, Malaysia

Dendrobium lanepoolei R. Rogers

Distribution: Papua New Guinea

Dendrobium langbianense Gagnep.

Distribution: Viet Nam

Dendrobium lanyaiae Seidenf.

Distribution: Thailand

Dendrobium lasianthera J.J.Sm.
Dendrobium ostrinoglossum Rupp
Dendrobium stueberi Hort.

Part II: Dendrobium

Distribution: Indonesia, Papua New Guinea

Dendrobium lasioglossum Rchb.f.
Callista lasioglossa (Rchb.f.) Kuntze

Distribution: Myanmar

Dendrobium laurensii J.J.Sm.
Sayeria laurensii (J.J.Sm.) Rauschert

Distribution: Indonesia

Dendrobium × lavarackianum M.A.Clem.
Dendrobium bigibbum var. *georgei* C.T.White
Dendrobium bigibbum var. *venosum* F.M.Bailey

Distribution: Australia

Dendrobium lawesii F.Muell.
Dendrobium pseudomohlianum Kraenzl.
Dendrobium salmonicolor Schltr.
Dendrobium warburgianum (Kraenzl.)
Pedilonum lawesii (F.Muell.) Rauschert

Distribution: Papua New Guinea, Solomon Islands

Dendrobium lawiense J.J.Sm.
Aporum lawiense (J.J.Sm.) Rauschert

Distribution: Malaysia

Dendrobium laxiflorum J.J.Sm.

Distribution: Indonesia

Dendrobium leonis (Lindl.) Rchb.f.
Aporum leonis Lindl.
Callista leonis (Lindl.) Kuntze
Dendrobium anceps auct. non Sw.

Distribution: Cambodia, Indonesia, Lao People's Democratic Republic, Malaysia, Thailand, Viet Nam

Dendrobium leporinum J.J.Sm.

Distribution: Indonesia

Dendrobium leptocladum Hayata

Distribution: China (Taiwan)

Dendrobium leucochlorum Rchb.f.
Callista leucochlora (Rchb.f.) Kuntze

Distribution: Myanmar

Dendrobium leucocyanum T.M.Reeve

Distribution: Papua New Guinea

Dendrobium leucohybos Schltr.
Dendrobium leucohybos var. *leucanthum* Schltr.
Latourorchis leucohybos (Schltr.) Brieger
Sayeria leucohybos (Schltr.) Rauschert

Distribution: Indonesia, Papua New Guinea

Dendrobium lichenastrum (F.Muell.) Nicholls
Bulbophyllum lichenastrum F.Muell.
Dockrillia lichenastrum (F.Muell.) Brieger
Phyllorchis lichenastrum (F.Muell.) Kuntze

Distribution: Australia

Dendrobium limii J.J.Wood

Distribution: Malaysia

Dendrobium linawianum Rchb.f.
Callista linawiana (Rchb.f.) Kuntze

Distribution: China, China (Taiwan)

Dendrobium lindleyi Steud.
Callista aggregata (Roxb.) Kuntze
Dendrobium suavissimum auct. non Rchb.f.
Dendrobium aggregatum Roxb.
Dendrobium jenkinsii auct. non Wall. ex. Lindl.

Distribution: Bhutan, China, India, Lao People's Democratic Republic, Myanmar, Thailand, Viet Nam

Dendrobium lineale Rolfe
Callista veratrifolia (Lindl.) Kuntze
Dendrobium augustae-victoriae Kraenzl.

137

Part II: Dendrobium

Dendrobium cogniauxianum Kraenzl.
Dendrobium grantii C.T.White
Dendrobium imperatrix Kraenzl.
Dendrobium veratrifolium Lindl.
Dendrobium veratroides Bakh.f.

Distribution: Papua New Guinea

Dendrobium linearifolium Teijsm. & Binn.
Dendrobium gracile Kraenzl. non Lindl.

Distribution: Indonesia

Dendrobium linguella Rchb.f.
Dendrobium hercoglossum auct. non Rchb.f.
Dendrobium poilanei auct. non Guillaum.

Distribution: Indonesia, Malaysia, Philippines (the), Thailand, Viet Nam

Dendrobium linguiforme Sw.
Callista linguiformis (Sw.) Kuntze
Dockrillia linguiforme (Sw.) Brieger

Distribution: Australia

Dendrobium lithocola D.L.Jones & M.A.Clem.
Dendrobium phalaenopsis var. *compactum* C.T.White

Distribution: Australia

Dendrobium litorale Schltr.
aporum litorale (Schltr.) Rauschert

Distribution: Australia, Papua New Guinea

Dendrobium lituiflorum Lindl.
Callista lituiflora (Lindl.) Kuntze
*Dendrobium hanburyanum*Rchb.f.
Dendrobium lituiflora (Lindl.) Kuntze

Distribution: China, India, Lao People's Democratic Republic, Myanmar, Thailand, Viet Nam

Dendrobium lobatum (Blume) Miq.
Aporum lobatum Blume
Callista lobata (Blume) Kuntze
Dendrobium rhizophoreti Ridl.

Distribution: Indonesia, Malaysia

Dendrobium lobulatum Rolfe & J.J.Sm.
Aporum lobulatum (Rolfe & J.J.Sm.) Brieger

Distribution: Brunei Darussalam, Indonesia, Malaysia

Dendrobium loddigesii Rolfe
Callista loddigesii (Rolfe) Kuntze
Dendrobium crepidatum auct. non Lindl.

Distribution: China, Lao People's Democratic Republic, Viet Nam

Dendrobium loesenerianum Schltr.
Pedilonum loesenerianum (Schltr.) Rauschert

Distribution: Papua New Guinea

Dendrobium lohoense T. Tang & F.T. Wang

Distribution: China

Dendrobium lomatochilum Seidenf.

Distribution: Cambodia, Viet Nam

Dendrobium longicornu Wall. ex Lindl.
Callista longicornis (Lindl.) Kuntze

Distribution: Bangladesh, Bhutan, China, India, Myanmar, Nepal, Viet Nam

Dendrobium lowii Lindl.
Callista lowii (Lindl.) Kuntze
Dendrobium lowii var. *pleiotrichum* Rchb.f.

Distribution: Indonesia, Malaysia

Dendrobium lucens Rchb.f.
Callista lucens (Rchb.f.) Kuntze

Distribution: Malaysia

Dendrobium lueckelianum Fessel & M.Wolff

Distribution: Thailand

Dendrobium lunatum Lindl.
Callista lunata (Lindl.) Kuntze

Part II: Dendrobium

Distribution: Philippines (the)

Dendrobium macarthiae Thw.
Callista macarthiae (Thw.) Kuntze

Distribution: Sri Lanka

Dendrobium macgregorii F.Muell. ex Kraenzl

Distribution: Philippines (the)

Dendrobium macranthum A.Rich.
Callista macrantha (A.Rich.) Kuntze
Dendrobium arachnostachyum Rchb.f.
Dendrobium pseudotokai Kraenzl.
Dendrobium tokai var. *crassinerve* Finet
Sayeria pseudotokai (Kraenzl.) Rauschert

Distribution: Solomon Islands, Vanuatu

Dendrobium macrifolium J.J.Sm.

Distribution: Indonesia

Dendrobium macrogenion Schltr.
Pedilonum macrogenion (Schltr.) Rauschert

Distribution: Papua New Guinea

Dendrobium macrophyllum A.Rich.
Callista gordonii (S.Moore) Kuntze
Callista veitchiana (Lindl.) Kuntze
Dendrobium brachythecum F.Muell. & Kraenzl.
Dendrobium ferox Hassk.
Dendrobium gordonii S.Moore
Dendrobium macrophyllum var. *veitchianum* (Lindl.) Hook.f.
Dendrobium musciferum Schltr.
Dendrobium psyche Kraenzl.
Dendrobium sarcostoma Teijsm. & Binn. ex Miq.
Dendrobium ternatense J.J.Sm.
Dendrobium tomohonense Kraenzl.
Dendrobium veitchianum Lindl.
Latourorchis macrophylla (A.Rich.) Brieger.
Latourorchis muscifera (Schltr.) Brieger
Sayeria macrophylla (A.Rich.) Rauschert
Sayeria muscifera (Schltr.) Rauschert

Distribution: Fiji, Indonesia, Papua New Guinea, Philippines (the), Samoa, Solomon Islands, United States of America (the), Vanuatu

140

Dendrobium macrophyllum var. **macrophyllum**

Distribution: Fiji, Indonesia, Papua New Guinea, Philippines (the), Samoa, Solomon Islands, United States of America (the), Vanuatu

Dendrobium macrophyllum var. **subvelutinum** J.J.Sm.

Distribution: Indonesia, Papua New Guinea

Dendrobium macropus (Endl.) Rchb.f. ex Lindl.
Callista macropus (Benth. & Bhhok.f.) Kuntze
Dendrobium comptonii Rendle
Dendrobium floribundum Rchb.f.
Dendrobium gracilicaule F.Muell.
Dendrobium gracilicaule var. *howeanum* Maiden
Dendrobium oscari A.D.Hawkes & A.H.Heller

Distribution: Australia, Fiji, New Caledonia, Samoa

Dendrobium macrostachyum Lindl.
Callista macrostachya (Lindl.) Kuntze
Dendrobium gamblei Duthie

Distribution: India, Sri Lanka

Dendrobium maierae J.J.Sm.

Distribution: Indonesia

Dendrobium magistratus P.J.Cribb

Distribution: Papua New Guinea, Solomon Islands

Dendrobium malvicolor Ridl.
Dendrobium hasseltii auct. non Lindl.
Dendrobium kuhlii auct. non Lindl.
Pedilonum malvicolor (Ridl.) Rauschert

Distribution: Indonesia

Dendrobium mannii Ridl.
Aporum mannii (Ridl.) Rauschert
Dendrobium nathanielis auct. non Rchb.f.
Dendrobium terminale auct. non Par. & Rchb.f.

Distribution: India, Lao People's Democratic Republic, Malaysia, Thailand, Viet Nam

Part II: Dendrobium

Dendrobium marmoratum Rchb.f.
Callista marmorata (Rchb.f.) Kuntze

Distribution: Myanmar

Dendrobium masarangense Schltr
Dendrobium pumilio Schltr.

Distribution: Fiji, Indonesia, Papua New Guinea, New Caledonia, Solomon Islands, Vanuatu

Dendrobium masarangense subsp. chlorinum Ridl.
Dendrobium chlorinum Ridl.
Pedilonum chlorinum (Ridl.) Rauschert

Distribution: Indonesia

Dendrobium masarangense subsp. masarangense
Pedilonum pumilio (Schltr.) Rauschert

Distribution: Fiji, Indonesia, Papua New Guinea, New Caledonia, Solomon Islands, Vanuatu

Dendrobium masarangense var. theionanthum (Schltr.) T.M.Reeve & P.Woods
Dendrobium caespitificum Ridl.
Dendrobium frigidum Schltr.
Dendrobium gemma Schltr.
Dendrobium monogrammoides J.J.Sm.
Dendrobium pseudofrigidum J.J.Sm.
Dendrobium theionanthum Schltr.
Pedilonum caespitificum (Ridl.) Rauschert
Pedilonum frigidum (Schltr.) Rauschert
Pedilonum gemma (Schltr.) Rauschert
Pedilonum theionanthum (Schltr.) Rauschert

Distribution: Indonesia, Papua New Guinea

Dendrobium mayandyi T.M.Reeve & Renz
Sayeria mayandyi (T.M.Reeve & Renz) Rauschert

Distribution: Papua New Guinea

Dendrobium melaleucaphilum M.A.Clem. & D.L.Jones

Distribution: Australia

Dendrobium melinanthum Schltr.
Pedilonum melinanthum (Schltr.) Rauschert

Distribution: Papua New Guinea

Dendrobium microbolbon A.Rich.
Callista microbolbon (A.Rich.) Kuntze
Dendrobium humile Wight

Distribution: Cambodia, India

Dendrobium militare P.J.Cribb
Dendrobium brevimentum P.J.Cribb non Seidenf.

Distribution: Indonesia

Dendrobium minimum Ames & C.Schweinf.

Distribution: Malaysia

Dendrobium minutiflorum S.C.Chen & Z.H. Tsi
Sayeria minutiflora (Gagnep.) Rauschert

Distribution: China, Japan

Dendrobium mirbelianum Gaudich.
Callista mirbeliana (Gaudich.) Kuntze
Dendrobium aruanum Kraenzl.
Dendrobium buluense Schltr.
Dendrobium buluense var. *kauloense* Schltr.
Dendrobium guilianettii F.M.Bailey
Dendrobium polycarpum Rchb.f.
Dendrobium rosenbergii Teijsm. & Binn.
Dendrobium wilkianum Rupp

Distribution: Australia, Indonesia, Papua New Guinea, Solomon Islands

Dendrobium miyakei Schltr.
Dendrobium hainanense Masam. & Hayata
Dendrobium pseudo-hainanense Masam
Dendrobium victoriae-reginae Loher var. miyakei (Schltr.) T.S.Liu & H.J.Su

Distribution: China (Taiwan)

Dendrobium modestissimum Kraenzl.

Distribution: Malaysia

Dendrobium modestum Rchb.f.

Distribution: Philippines (the)

Dendrobium mohlianum Rchb.f.
Dendrobium neo-ebudanum Schltr.
Dendrobium vitellinum Kraenzl.
Pedilonum mohlianum (Rchb.f.) Brieger
Pedilonum neoebudanum (Schltr.) Rauschert
Pedilonum vitellinum (Kraenzl.) Rauschert

Distribution: Fiji, Samoa, Solomon Islands, Vanuatu

Dendrobium molle J.J.Sm.
Pedilonum molle (J.J.Sm.) Rauschert

Distribution: Indonesia, Papua New Guinea

Dendrobium moniliforme (L.) Sw.
Callista japonica (Blume) Kuntze
Callista moniliformis Rolfe
Dendrobium castum Bateman ex Hook.f.
Dendrobium catenatum Lindl.
Dendrobium japonicum Lindl.
Dendrobium monile (Thunb.) Kraenzl.
Dendrobium zonatum Rolfe
Epidendrum monile Thunb.
Epidendrum moniliferum Panzer
Epidendrum moniliforme L.
Onychium japonicum Blume
Ormostema purpurea Raf.
Ormostemam albiflora Raf.

Distribution: China, China (Taiwan), Japan, Korea (the Republic of)

Dendrobium monophyllum F.Muell.
Australorchis monophylla (F.Muell.) Brieger
Callista monophylla (F.Muell.) Kuntze
Dendrobium tortile A.Cunn.

Distribution: Australia

Dendrobium montanum J.J.Sm.

Distribution: Indonesia

Dendrobium monticola Hunt & Summerh.
Callista alpestris (Royle) Kuntze
Dendrobium alpestre Royle
Dendrobium denudans auct. non D.Don & Lindl.
Dendrobium pusillum D.Don
Dendrobium roylei Hawkes & Heller
Dendrobium roylei Hiroe

Distribution: China, India, Nepal, Thailand

Dendrobium montis-yulei Kraenzl.

Distribution: Papua New Guinea

Dendrobium mooreanum Lindl.
Dendrobium fairfaxii Rolfe
Dendrobium petri Rchb.f.
Dendrobium priscillae A.D.Hawkes
Dendrobium quaifei Rolfe ex Ames
Sayeria mooreana (Lindl.) Rauschert

Distribution: Vanuatu

Dendrobium moorei F.Muell.
Callista moorei (F. Muell.) Kuntze
Tropilis moorei (F.Muell.) Rauschert

Distribution: Australia

Dendrobium mortii F.Muell.
Callista mortii (F.Muell.) Kuntze
Dendrobium robertsii F.Muell. ex Rupp
Dendrobium tenuissimum Rupp
Dockrillia mortii (F.Muell.) Rauschert
Dockrillia tenuissima (Rupp) Rauschert

Distribution: Australia

Dendrobium moschatum (Buch.-Ham.) Sw.
Callista moschata (Buch.-Ham.) Kuntze
Cymbidium moschatum (Buch.-Ham.) Wild.
Dendrobium calceolaria Carey ex Hook.
Dendrobium cupreum Herbert
Epidendrum moschatum Buch.-Ham.

Distribution: Bhutan, China, India, Lao People's Democratic Republic, Myanmar, Nepal,
Thailand, Viet Nam

Dendrobium moulmeinense Rchb.f.
Callista moulmeinensis (Par. & Rchb.f.) Kuntze

Distribution: Myanmar

Dendrobium mucronatum Seidenf.

Distribution: Thailand

Part II: Dendrobium

Dendrobium multilineatum Kerr

Distribution: Lao People's Democratic Republic

Dendrobium multiramosum Ames

Distribution: Philippines (the)

Dendrobium mutabile (Blume) Lindl.
  Callista mutabilis (Blume) Kuntze
  Onychium mutabile Blume

Distribution: Indonesia

Dendrobium mystroglossum Schltr.
  Pedilonum mystroglossum (Schltr.) Rauschert

Distribution: Papua New Guinea

Dendrobium nardoides Schltr.
  Dendrobium oligoblepharon Schltr.
  Pedilonum nardoides (Schltr.) Rauschert
  Pedilonum oligoblepharon (Schltr.) Rauschert

Distribution: Papua New Guinea

Dendrobium nathanielis Rchb.f.
  Aporum anceps (Sw.) Lindl.
  Aporum cuspidatum Wall. ex Lindl.
  Callista nathanielis (Rchb.f.) Kuntze
  Dendrobium anceps auct. non Sw.
  Dendrobium cuspidatum auct. non Lindl.
  Dendrobium mannii auct. non Ridl.
  Dendrobium multiflorum Parish & Rchb.f.

Distribution: Cambodia, India, Lao People's Democratic Republic, Myanmar, Thailand, Viet Nam

Dendrobium navicula Kraenzl.
  Pedilonum navicula (Kraenzl.) Rauschert

Distribution: Papua New Guinea

Dendrobium nebularum Schltr.
  Dendrobium keysseri Schltr.
  Dendrobium murkelense J.J.Sm.
  Dendrobium palustre L.O.Williams
  Dendrobium tumidulum Schltr.
  Pedilonum keysseri (Schltr.) Rauschert

146

Pedilonum murkelense (J.J.Sm.) Rauschert
Pedilonum nebularum (Schltr.) Rauschert
Pedilonum tumidulum (Schltr.) Rauschert

Distribution: Indonesia, Papua New Guinea

Dendrobium nieuwenhuisii J.J.Sm.

Distribution: Indonesia

Dendrobium nindii W.Hill
Dendrobium ionoglossum Schltr.
Dendrobium ionoglossum var. *potamophilum* Schltr.
Dendrobium tofftii F.M.Bailey

Distribution: Australia, Papua New Guinea

Dendrobium nobile Lindl.
Callista nobilis (Lindl.) Kuntze
Dendrobium chlorostylum Gagnep.
Dendrobium coerulescens Wall.
Dendrobium formosanum (Rchb.f.) Masamune
Dendrobium friedericksianum auct. non Rchb.f.
Dendrobium lindleyanum Griff.
Dendrobium nobile var. *formosanum* Rchb.f.

Distribution: Bhutan, China, China (Taiwan), India, Lao People's Democratic Republic, Myanmar, Nepal, Thailand, Viet Nam

Dendrobium nobile var. **alboluteum** Huyen & Aver.

Distribution: Viet Nam

Dendrobium nobile var. **nobile**

Distribution: Bhutan, China, China (Taiwan), India, Lao People's Democratic Republic, Myanmar, Nepal, Thailand, Viet Nam

Dendrobium nothofagicola T.M.Reeve

Distribution: Papua New Guinea

Dendrobium nubigenum Schltr.
Pedilonum nubigenum (Schltr.) Rauschert

Distribution: Papua New Guinea

Dendrobium nudum (Blume) Lindl.

Part II: Dendrobium

Callista aurorosea (Rchb.f.) Kuntze
Callista nuda (Blume) Kuntze
Dendrobium auroroseum Rchb.f.
Onychium nudum Blume
Pedilonum auroroseum (Lindl.) Rauschert

Distribution: Indonesia

Dendrobium nugentii (F.M.Bailey) D.Jones & M.Clements
Dendrobium linguiforme var. *nugentii* F.M.Bailey
Dendrobium obcuneatum F.M.Bailey
Dockrillia nugentii (F.M.Bailey) M.A.Clem. & D.L.Jones

Distribution: Australia

Dendrobium nycteridoglossum Rchb.f.
Aporum nycteridoglossum (Rchb.f.) Rauschert
Aporum platyphyllum (Schltr.) Rauschert
Dendrobium platyphyllum Schltr.

Distribution: Malaysia

Dendrobium obcordatum J.J.Sm.
Aporum obcordatum (J.J.Sm.) Rauschert

Distribution: Indonesia

Dendrobium obrienianum Kraenzl.

Distribution: Philippines (the)

Dendrobium obtusum Schltr.
Dendrobium concavissimum J.J.Sm.
Dendrobium chrysoglossum Schltr.
Dendrobium fornicatum Schltr.
Dendrobium lauterbachianum A.D.Hawkes
Dendrobium rhodobotrys Ridl.
Pedilonum chrysoglossum (Schltr.) Rauschert
Pedilonum concavissimum (J.J.Sm.) Rauschert
Pedilonum fornicatum (Schltr.) Rauschert
Pedilonum obtusum (Schltr.) Rauschert
Pedilonum rhodobotrys (Ridl.) Rauschert

Distribution: Indonesia, Papua New Guinea

Dendrobium ochraceum De Wild.

Distribution: Viet Nam

Dendrobium ochreatum Lindl.
Callista ochreata (Lindl.) Kuntze
Dendrobium cambridgeanum Paxton

Distribution: Bhutan, China, India, Lao People's Democratic Republic, Myanmar, Thailand, Viet Nam

Dendrobium odoardii Kraenzl.

Distribution: Indonesia

Dendrobium officinale Kimura & Migo

Distribution: China

Dendrobium okinawense Hatusima

Distribution: Japan

Dendrobium oliganthum Schltr.

Distribution: Indonesia

Dendrobium oligophyllum Gagnep.
Dendrobium Tixieri Guill.

Distribution: Thailand, Viet Nam

Dendrobium oreodoxa Schltr.
Pedilonum oreodoxa (Schltr.) Rauschert

Distribution: Papua New Guinea

Dendrobium oreogenum Schltr.

Distribution: Indonesia, Papua New Guinea

Dendrobium otaguroanum A.D.Hawkes
Dendrobium chloroleucum Schltr.
Sayeria otaguroana (A.D. Hawkes) Rauschert

Distribution: Papua New Guinea

Dendrobium ovatum (Willd.) Kraenzl.
Cymbidium ovatum Willd.
Dendrobium barbatulum auct. non Lindl.

Part II: Dendrobium

Dendrobium chlorops Lindl.

Distribution: India

Dendrobium ovipostoriferum J.J.Sm.

Distribution: Indonesia, Malaysia

Dendrobium paathii J.J.Sm.

Distribution: Indonesia

Dendrobium pachyglossum Par. & Rchb.f.
Callista pachyglossa (Par. & Rchb.f.) Kuntze
Dendrobium fallax Guill.

Distribution: Lao People's Democratic Republic, Myanmar, Thailand, Viet Nam

Dendrobium pachystele Schltr.
Dendrobium pachystele var. *homeoglossum* Schltr.
Sayeria pachystele (Schltr.)Rauschert

Distribution: Papua New Guinea, Solomon Islands, Palau, Micronesia (Federated States of)

Dendrobium pachythrix T.M.Reeve & P.Woods

Distribution: Papua New Guinea

Dendrobium palpebrae Lindl.
Callista palpebrae (Lindl.) Kuntze
Dendrobium densiflorum auct. non Lindl.
Dendrobium farmeri auct. non Paxt.
Dendrobium farmeri var. *album* Regel

Distribution: India, Lao People's Democratic Republic, Myanmar, Thailand, Viet Nam

Dendrobium panduriferum Hook.f.
Callista pandurifera (Hook.f.) Kuntze
Dendrobium ionopus auct. non Rchb.f.
Dendrobium panduriferum var. *serpens* Hook.f.
Dendrobium serpens (Hook.f.) Hook.f.
Dendrobium virescens Ridl.
Pedilonum panduriferum (Hook.f.) Brieger
Pedilonum serpens (Hook.f.) Brieger

Distribution: Malaysia, Myanmar, Thailand

Dendrobium paniferum J.J.Sm.

Distribution: Indonesia

Dendrobium papilio Loher
Aporum papilio (Loher) Rauschert

Distribution: Philippines (the)

Dendrobium papilioniferum J.J.Sm.
Angraecum crumenatum Rumph.
Dendrobium crumenatum var. *papilioniferum* Kraenzl.

Distribution: Indonesia

Dendrobium papilioniferum var. **ephemerum** J.J.Sm.
Angraecum album-minus Rumph.
Aporum ephemerum (J.J.Sm.) Rauschert
Dendrobium ephemerum J.J.Sm.

Distribution: Indonesia

Dendrobium papilioniferum var. **papilioniferum**

Distribution: Indonesia

Dendrobium papuanum J.J.Sm.
Pedilonum papuanum (J.J.Sm.) Rauschert

Distribution: Indonesia

Dendrobium parciflorum Rchb. ex Lindl.
Aporum jenkinsii Griff.
Callista jenkinsii (Griff.) Kuntze

Distribution: China, India, Lao People's Democratic Republic, Thailand, Viet Nam

Dendrobium parcum Rchb.f.
Callista parca (Rchb.f.) Kraenzl.
Dendrobium listeroglossum Kraenzl.
Dendrobium parcoides Guillaum.

Distribution: Myanmar, Thailand, Viet Nam

Dendrobium parishii Rchb.f.
Callista parishii (Rchb.f.) Kuntze

Distribution: China, India, Lao People's Democratic Republic, Myanmar, Thailand, Viet Nam

Part II: Dendrobium

Dendrobium parthenium Rchb.f.

Distribution: Malaysia

Dendrobium parvifolium J.J.Sm.
Pedilonum parvifolium (J.J.Sm.) Rauschert

Distribution: Indonesia

Dendrobium paspalifolium J.J.Sm.

Distribution: Indonesia

Dendrobium patentilobum Ames & C.Schweinf.

Distribution: Malaysia

Dendrobium pauciflorum King & Pantl.
Dendrobium sikkimense Hawkes & Heller

Distribution: Bhutan, India, Myanmar, Thailand

Dendrobium pedicellatum J.J.Sm.

Distribution: Indonesia

Dendrobium pedunculatum (Clemesha) D.L.Jones & M.A.Clem.
Dendrobium speciosum var. *pedunculatum* Clemesha

Distribution: Australia

Dendrobium peguanum Lindl.
Callista pygmaea (Lindl.) Kuntze
Dendrobium microbolbon auct. non A.Rich.
Dendrobium pygmaeum Lindl.
Dendrobium wallichii Hawkes & Heller

Distribution: India, Myanmar, Nepal

Dendrobium pendulum Roxb.
Callista crassinodis (Benson & Rchb.f.) Kuntze
Callista pendula (Roxb.) Kuntze
Dendrobium crassinode Benson & Rchb.f.
*Dendrobium melanophthalmum*Rchb.f.

Distribution: China, India, Lao People's Democratic Republiclic, Myanmar, Thailand, Viet Nam

Dendrobium pentapterum Schltr.
Pedilonum pentapterum (Schltr.) Rauschert

Distribution: Papua New Guinea

Dendrobium percnanthum Rchb.f.

Distribution: Indonesia

Dendrobium perulatum Gagnep.

Distribution: Viet Nam

Dendrobium petiolatum Schltr.
Dendrobium unifoliatum Schltr.
Pedilonum petiolatum (Schltr.) Rauschert

Distribution: Papua New Guinea, Solomon Islands

Dendrobium phalaenopsis Fitzg.
Callista phalaenopsis (Fitzg.) Kuntze
Dendrobium bigibbum var. *phalaenopsis* (Fitzg.) Bailey
Dendrobium bigibbum var. *superbum* Hort. ex Rchb.f.

Distribution: Australia

Dendrobium pictum Lindl.

Distribution: Malaysia

Dendrobium planibulbe Lindl.
Aporum planibulbe (Lindl.) Rauschert
Callista tuberifera (Hook.f.) Kuntze
Dendrobium blumei auct. non Lindl.
Dendrobium tuberiferum Hook.f.

Distribution: Indonesia, Malaysia, Thailand

Dendrobium pleurodes Schltr.
Sayeria pleurodes (Schltr.) Rauschert

Distribution: Papua New Guinea

Dendrobium podagraria Hook.f.
Callista angulata (Lindl.) Kuntze
Dendrobium angulatum Wall. ex Lindl.
Dendrobium inconcinnum Ridl.

Distribution: India, Myanmar, Thailand, Viet Nam

Dendrobium pogoniates Rchb.f.

Distribution: Indonesia, Malaysia

Dendrobium polysema Schltr.
Dendrobium macrophyllum var. *stenopterum* Rchb.f.
Dendrobium polysema var. *pallidum* Chadim
Dendrobium pulchrum Schltr.
Sayeria polysema (Schltr.) Rauschert

Distribution: Papua New Guinea, Solomon Islands, Vanuatu

Dendrobium porphyrochilum Lindl.
Callista porphyrochila (Lindl.) Kuntze
Dendrobium caespitosum King & Pantl.
Dendrobium confinale auct. non Kerr

Distribution: Bhutan, China, India, Myanmar, Nepal

Dendrobium porphyrophyllum Guill.
Dendrobium indivisum var. *lampangense* Rolfe
Dendrobium neolampangense Aver.

Distribution: Lao People's Democratic Republic, Thailand, Viet Nam

Dendrobium praetermissum Dauncey
Dendrobium dichaeoides auct. non Schltr.

Distribution: Indonesia, Papua New Guinea

Dendrobium prasinum Lindl.
Sarcopodium prasinum (Lindl.) Kraenzl.

Distribution: Fiji

Dendrobium prenticei (F.Muell.) Nicholls
Bulbophyllum lichenastrum auct. non. F.Muell.
Bulbophyllum prenticei F.Muell.
Dendrobium aurantiaco-purpureum Nicholls
Dendrobium lichenastrum var. *prenticei* (F.Muell.) Dockrill
Dendrobium variabile Nicholls

Distribution: Australia

Dendrobium prianganense J.J.Wood

Distribution: Indonesia

Dendrobium primulinum Lindl.
Callista primulina (Lindl.) Kuntze
Dendrobium nobile var. *pallidiflora* Hook.

Distribution: China, India, Lao People's Democratic Republic, Myanmar, Nepal, Thailand, Viet Nam

Dendrobium pristinum Ames

Distribution: Philippines (the)

Dendrobium profusum Rchb.f.

Distribution: Philippines (the)

Dendrobium prostratum Ridl.
Aporum prostratum (Ridl.) Rauschert

Distribution: Malaysia, Singapore

Dendrobium proteranthum Seidenf.

Distribution: Thailand

Dendrobium pseudoaloifolium J.J.Wood

Distribution: Malaysia

Dendrobium pseudocalceolum J.J.Sm.
Aporum pseudocalceolum (J.J.Sm.) Rauschert
Dendrobium cuspidatum Kraenzl.

Distribution: Papua New Guinea

Dendrobium pseudoconanthum J.J.Sm.

Distribution: Indonesia

Dendrobium pseudoglomeratum T.M.Reeve & J.J.Woods
Dendrobium chrysoglossum auct. non Schltr.
Dendrobium glomeratum auct. non Rolfe

Distribution: Indonesia, Papua New Guinea

Part II: Dendrobium

Dendrobium pseudointricatum Guillaum.

Distribution: Viet Nam

Dendrobium pseudopeloricum J.J.Sm.

Distribution: Indonesia, Papua New Guinea

Dendrobium pseudorarum Dauncey
Dendrobium rarum auct. non Schltr.
Dendrobium calcaratum auct. non A.Rich

Distribution: Vanuatu

Dendrobium pseudorarum var. **baciforme** Dauncey
Dendrobium rarum auct. non Schltr.
Dendrobium calcaratum auct. non A.Rich

Distribution: Vanuatu

Dendrobium pseudorarum var. **pseudorarum**

Distribution: Vanuatu

Dendrobium pseudotenellum Guillaum.
Dendrobium tenellum auct. non Lindl.
Dendrobium tenellum var. *setifolium* Guillaum.

Distribution: China, Viet Nam

Dendrobium puberilingue J.J.Sm.

Distribution: Indonesia, Malaysia

Dendrobium pugioniforme A.Cunn.
Callista pugioniformis (A.Cunn.) Kuntze
Dendrobium pugentifolium F.Muell.
Dockrillia pugioniformis (A.Cunn.) Rauschert

Distribution: Australia

Dendrobium pulchellum Roxb. ex Lindl.
Callista pulchella (Roxb. ex Lindl.) Kuntze
Dendrobium dalhousieanum Wall.

Distribution: China, India, Lao People's Democratic Republic, Malaysia, Myanmar, Nepal, Thailand, Viet Nam

Dendrobium punamense Schltr.
Dendrobium waterhousei Carr
Sayeria punamensis (Schltr.) Rauschert

Distribution: Papua New Guinea, Solomon Islands

Dendrobium puniceum Ridl.
Dendrobium adolphi Schltr.
Dendrobium cerasinium Ridl.
Dendrobium discrepans J.J.Sm.
Dendrobium lateriflorum Ridl.
Dendrobium scarlatinum Schltr.
Dendrobium subacaule sensu Kraenzl.
Pedilonum adolphi (Schltr.) Rauschert
Pedilonum cerasinum (Ridl.) Rauschert
Pedilonum discrepans (J.J.Sm.) Rauschert
Pedilonum lateriflorum (Ridl.) Rauschert
Pedilonum puniceum (Schltr.) Rauschert
Pedilonum scarlatinum (Schltr.) Rauschert

Distribution: Papua New Guinea, Solomon Islands

Dendrobium purpureiflorum J.J.Sm.
Pedilonum purpureiflorum (J.J.Sm.) Rauschert

Distribution: Indonesia

Dendrobium purpureum Roxb.
Angraecum purpureum var. *sylvestre* Rumph.
Callista purpurea (Roxb.) Kuntze
Callista reinwardtii (Lindl.) Kuntze
Dendrobium moseleyi Hemsl.
Dendrobium praeustum Kraenzl.
Dendrobium purpureum var. *album* Hort.
Dendrobium purpureum var. *moseleyi* Hemsl.
Dendrobium reinwardtii Lindl.
Dendrobium scabripes Kraenzl.
Dendrobium viridiroseum Rchb.f.
Pedilonum purpureum (Roxb.) Brieger
Pedilonum reinwardtii (Lindl.) Rauschert

Distribution: Indonesia

Dendrobium purpureum subsp. **candidulum** (Rchb.f.) Dauncey & P.J.Cribb
Dendrobium purpureum var. *candidulum* Rchb.f.
Dendrobium purpureum var. *steffensianum* Schltr.
Dendrobium talaudense J.J.Sm.
Dendrobium viridiroseum var. *candidulum* Rchb.f.
Pedilonum talaudense (J.J.Sm.) Rauschert

Distribution: Indonesia

Part II: Dendrobium

Dendrobium purpureum subsp. **purpureum**

Distribution: Indonesia

Dendrobium putnamii Hawkes & Heller
Dendrobium coerulescens Schltr.
Pedilonum putnamii (Hawkes & Heller) Rauschert

Distribution: Papua New Guinea

Dendrobium pychnostachyum Lindl.
Callista pychnostachya (Lindl.) Kuntze

Distribution: India, Myanmar, Thailand

Dendrobium quadriquetrum J.J.Sm.
Pedilonum quadriquetrum (J.J.Sm.) Rauschert

Distribution: Indonesia

Dendrobium racemosum (Nicholls) Clemesha & Dockrill
Dendrobium beckleri var. *racemosum* Nicholls
Dockrillia racemosa (Nicholls) Rauschert

Distribution: Australia

Dendrobium rachmatii J.J.Sm.

Distribution: Indonesia

Dendrobium radians Rchb.f.
Callista radians (Rchb.f.) Kuntze

Distribution: Malaysia

Dendrobium ramosii Ames
Callista ramosa (Lindl.) Kuntze

Distribution: Philippines (the)

Dendrobium rantii J.J.Sm.

Distribution: Indonesia

Dendrobium rappardii J.J.Sm.

Distribution: Indonesia

Dendrobium rariflorum J.J.Sm.

Distribution: Indonesia

Dendrobium rarum Schltr.
  Pedilonum rarum (Schltr.) Rauschert

Distribution: Indonesia, Papua New Guinea

Dendrobium rarum var. **miscegeneum** Dauncey

Distribution: Indonesia, Papua New Guinea

Dendrobium rarum var. **pelorium** Dauncey

Distribution: Papua New Guinea

Dendrobium rarum var. **rarum**

Distribution: Indonesia, Papua New Guinea

Dendrobium reflexibarbatulum J.J.Sm.

Distribution: Indonesia

Dendrobium reflexitepalum J.J.Sm.
  Aporum reflexitepalum (J.J.Sm.) Rauschert

Distribution: Indonesia

Dendrobium rennellii P.J.Cribb

Distribution: Solomon Islands

Dendrobium rex M.A.Clem. & D.L.Jones
  Dendrobium speciosum var. *grandiflorum* F.M.Bailey

Distribution: Australia

Dendrobium rhabdoglossum Schltr.
Pedilonum rhabdoglossum (Schltr.) Rauschert

Distribution: Papua New Guinea

Part II: Dendrobium

Dendrobium rhodopterygium Rchb.f.
Callista rhodopterygia (Rchb.f.) Kuntze
Dendrobium polyphlebium Rchb.f.

Distribution: Myanmar

Dendrobium rhodostele Ridl.
Aporum rhodostele (Ridl.) Rauschert

Distribution: Indonesia, Malaysia, Thailand

Dendrobium rhodostictum F.Muell. & Kraenzl.
Dendrobium madonnae Rolfe
Sayeria rhodosticta (F.Muell. & Kraenzl.) Rauschert

Distribution: Papua New Guinea, Solomon Islands

Dendrobium rhomboglossum J.J.Sm.
Sayeria rhomboglossa (J.J.Sm.) Rauschert

Distribution: Indonesia

Dendrobium rigidifolium Rolfe
Dendrobium alpinum P.Royen
Dendrobium giluwense P.Royen
Dendrobium guttatum J.J.Sm.
Dendrobium helenae Chadim
Sayeria rigidifolia (Rolfe) Rauschert

Distribution: Indonesia, Papua New Guinea

Dendrobium rigidum R.Br.
Callista rigida (R.Br.) Kuntze
Dendrobium desmotrichoides J.J.Sm.
Dockrillia rigida (R.Br.) Rauschert
Dockrillia desmotrichoides (J.J.Sm.) Brieger

Distribution: Australia, Indonesia, Papua New Guinea

Dendrobium rindjaniense J.J.Sm.
Pedilonum rindjaniense (J.J.Sm.) Rauschert

Distribution: Indonesia

Dendrobium riparium J.J.Sm.
Pedilonum riparium (J.J.Sm.) Rauschert

Distribution: Indonesia

Dendrobium roseicolor A.D. Hawkes & A.H. Heller
Dendrobium roseum Schltr. non Sw.
Pedilonum roseicolor (A.D. Hawkes & A.H. Heller) Rauschert
Pedilonum roseum (Schltr.) Brieger

Distribution: Papua New Guinea

Dendrobium roseipes Schltr.
Pedilonum roseipes (Schltr.) Rauschert

Distribution: Papua New Guinea

Dendrobium rosellum Ridl.
Aporum rosellum (Ridl.) Rauschert

Distribution: Brunei Darussalam, Indonesia, Malaysia

Dendrobium ruginosum Ames
Sayeria ruginosa (Ames) Rauschert

Distribution: Solomon Islands

Dendrobium rupestre J.J.Sm.
Pedilonum rupestre (J.J.Sm.) Rauschert

Distribution: Indonesia, Papua New Guinea

Dendrobium × ruppiosum Clemesha

Distribution: Australia

Dendrobium rutriferum Rchb.f.
Pedilonum rutriferum (Rchb.f.) Rauschert

Distribution: Indonesia, Papua New Guinea

Dendrobium ruttenii J.J.Sm.
Sayeria ruttenii (J.J.Sm.) Rauschert

Distribution: Indonesia

Dendrobium sagittatum J.J.Sm.
Aporum sagittatum (J.J.Sm.) Rauschert

Distribution: Indonesia

Part II: Dendrobium

Dendrobium salmoneum Schltr.
Pedilonum salmoneum (Schltr.) Rauschert

Distribution: Indonesia, Papua New Guinea

Dendrobium sambasanum J.J.Sm.

Distribution: Indonesia

Dendrobium samoense P.J.Cribb

Distribution: Samoa

Dendrobium sancristobalense P.J.Cribb

Distribution: Solomon Islands

Dendrobium sanderae Rolfe

Distribution: Philippines (the)

Dendrobium sanderianum Rolfe

Distribution: Malaysia

Dendrobium sanguinolentum Lindl.
Callista sanguinolenta (Lindl.) Kuntze
Dendrobium cerinum Rchb.f.
Pedilonum sanguinolentum (Lindl.) Brieger

Distribution: Indonesia, Malaysia, Philippines (the), Thailand

Dendrobium sarawakense Ames
Dendrobium multiflorum Ridl.

Distribution: Malaysia

Dendrobium scabrifolium Ridl.
Pedilonum scabrifolium (Ridl.) Rauschert

Distribution: Indonesia

Dendrobium scabrilingue Lindl.
Callista scabrillinguis (Lindl.) Kuntze
Dendrobium galactanthum Schlechter
Dendrobium hedyosmum Bateman

Distribution: Lao People's Democratic Republic, Myanmar, Thailand

Dendrobium schneiderae F.M.Bailey
  Australorchis schneiderae (F.M.Bailey) Brieger

Distribution: Australia

Dendrobium schneiderae var. **major** Rupp

Distribution: Australia

Dendrobium schneiderae var. **schneiderae**

Distribution: Australia

Dendrobium schoeninum Lindl.
  Callista beckleri (F.Muell.) Kuntze
  Dendrobium beckleri F.Muell.
  Dendrobium mortii Benth.
  Dendrobium striolatum F.M.Bailey
  Dendrobium striolatum var. *beckleri* (F.Muell.) F.M.Bailey
  Dockrillia beckleri (F.Muell.) Rauschert
  Dockrillia schoenina (Lindl.) M.A.Clem. & D.L.Jones

Distribution: Australia

Dendrobium schroederi Rolfe

Distribution: Philippines (the)

Dendrobium schuetzei Rolfe

Distribution: Philippines (the)

Dendrobium schulleri J.J.Sm.

Distribution: Indonesia

Dendrobium × **schumannianum** Schltr.

Distribution: Papua New Guinea

Dendrobium sculptum Rchb.f.
  Callista sculpta (Rchb.f.) Kuntze

Distribution: Malaysia

Dendrobium secundum (Blume) Lindl. ex Wall.
Callista bursigera (Lindl.) Kuntze
Callista secunda (Blume) Kuntze
Dendrobium bursigerum Lindl.
Dendrobium heterostigma Rchb.f.
Dendrobium secundum var. *bursigerum* (Lindl.) Ridl.
Dendrobium secundum var. *niveum* Rchb.f.
Pedilonum bursigerum (Lindl.) Rauschert
Pedilonum secundum Blume

Distribution: Brunei Darussalam, Cambodia, India, Indonesia, Lao People's Democratic Republic, Malaysia, Myanmar, Philippines (the), Singapore, Thailand, Viet Nam

Dendrobium senile Parish & Rchb.f.
Callista senilis (Par & Rchb.f.) Kuntze

Distribution: Lao People's Democratic Republic, Myanmar, Thailand

Dendrobium seranicum J.J.Sm.
Pedilonum seranicum (J.J.Sm.) Rauschert

Distribution: Indonesia

Dendrobium setifolium Ridl.
Aporum setifolium (Ridl.) Rauschert
Aporum peculiare (J.J.Sm.) Rauschert
Dendrobium gracile auct. non Lindl.
Dendrobium peculiare auct. non J.J.Sm.

Distribution: Indonesia, Malaysia, Singapore, Thailand

Dendrobium sidikalangense Dauncey

Distribution: Indonesia

Dendrobium signatum Rchb.f.
Dendrobium hildebrandii Rolfe
Dendrobium tortile var. *hildebrandii* (Rolfe) Tang & Wang

Distribution: Lao People's Democratic Republic, Myanmar, Thailand

Dendrobium simondii Gagnep.

Distribution: Viet Nam

Dendrobium simplex J.J.Sm.

Dendrobium sarcopodioides J.J.Sm.
Epigeneium simplex (J.J.Sm.) Summerh.
Katherinea simplex (J.J.Sm.) A.D.Hawkes
Sayeria simplex (J.J.Sm.) Rauschert

Distribution: Indonesia, Papua New Guinea

Dendrobium sinense Tang & Wang

Distribution: China

Dendrobium singkawangense J.J.Sm.

Distribution: Indonesia, Malaysia

Dendrobium sinuosum Ames

Distribution: Philippines (the)

Dendrobium smillieae F.Muell.
Callista ophioglossa (Rchb.f.) Kuntze
Callista smillieae (F.Muell.) Kuntze
Coelandria smillieae (F.Muell.) Fitzg.
Dendrobium hollrungii Kraenzl.
Dendrobium hollrungii var. *australiense* Rendle
Dendrobium kaernbachii Kraenzl.
Dendrobium ophioglossum Rchb.f.
Dendrobium pachyceras F.Muell. & Kraenzl.
Dendrobium secundum var. *urvillei* Finet
Dendrobium smillieae var. *hollrungii* (Kraenzl.) J.J.Sm.
Dendrobium smillieae var. *ophiglossum* (Rchb.f.) F.M.Bailey
Pedilonum hollrungii (Kraenzl.) Rauschert
Pedilonum ophioglossum (Rchb.f.) Brieger
Pedilonum pachyceras (F.Muell. & Kraenzl.) Rauschert
Pedilonum smillieae (F.Muell.) Rauschert

Distribution: Australia, Indonesia, Papua New Guinea

Dendrobium smithianum Schltr.
Aporum smithianum (Schltr.) Rauschert

Distribution: Indonesia, Malaysia

Dendrobium somai Hayata

Distribution: China (Taiwan)

Dendrobium soriense Howcroft

Part II: Dendrobium

Distribution: Papua New Guinea

Dendrobium spathilingue J.J.Sm.

Distribution: Indonesia

Dendrobium spathulatum L.O.Williams

Distribution: Fiji

Dendrobium speciosum Sm.
  Callista speciosa (Sm.) Kuntze

Distribution: Australia, Papua New Guinea

Dendrobium spectabile (Blume) Miq.
  Callista spectabilis (Blume) Kuntze
  Dendrobium tigrinum Rolfe ex Hemsl.
  Latourea spectabilis Blume
  Latourorchis spectabile (Blume) Brieger
  Tropilis speciosa (Sm.) Butzin
  Sayeria spectabilis (Blume) Rauschert

Distribution: Indonesia, Papua New Guinea, Solomon Islands

Dendrobium spectatissimum Rchb.f.
  Dendrobium reticulatum J.J.Sm.
  Dendrobium speciosissimum Rolfe

Distribution: Malaysia

Dendrobium spegidoglossum Rchb.f.
  Callista flavidula (Ridl. ex Hook.f.) Kuntze
  Dendrobium exsculptum Teijsm. & Binn.
  Dendrobium flavidulum Ridl. ex Hook.f.
  Dendrobium stuposum auct. non Lindl.

Distribution: Indonesia, Malaysia, Myanmar, Singapore, Thailand

Dendrobium stellare Dauncey

Distribution: Indonesia

Dendrobium stolleanum Schltr.
Pedilonum stolleanum (Schltr.) Rauschert

Distribution: Papua New Guinea

166

Dendrobium stratiotes Rchb.f.
Callista stratiotes (Rchb.f.) Kuntze
Dendrobium strebloceras var. *rossianum* Rchb.f.

Distribution: Indonesia

Dendrobium strebloceras Rchb.f.
Callista strebloceras (Rchb.f.) Kuntze
Dendrobium dammerboeri J.J.Sm.

Distribution: Indonesia

Dendrobium strepsiceros J.J.Sm.

Distribution: Indonesia

Dendrobium striolatum Rchb.f.
Callista striolata (Rchb.f.) Kuntze
Dendrobium milliganii F.Muell.
Dendrobium teretifolium Lindl.
Dockrillia striolata (Rchb.f.) Rauschert

Distribution: Australia

Dendrobium strongylanthum Rchb.f.
Callista strongylantha (Rchb.f.) Kuntze
Dendrobium ctenoglossum Schltr.

Distribution: China, Myanmar, Thailand

Dendrobium stuartii F.M.Bailey
Callista stuartii (F.M.Bailey) Kuntze
Dendrobium tetrodon auct. non Rchb.f.
Dendrobium whiteanum T.E.Hunt

Distribution: Australia, Indonesia, Malaysia, Papua New Guinea, Thailand, Viet Nam

Dendrobium stuposum Lindl.
Callista stuposa (Lindl.) Kuntze
Dendrobium spegidoglossum auct. non Rchb.f.

Distribution: Bhutan, China, India, Myanmar, Thailand

Dendrobium subacaule Reinw. ex Lindl.
Callista subacaulis (Reinw. ex Lindl.) Kuntze
Dendrobium begoniicarpum J.J.Sm.
Dendrobium begoniicarpum var. *parviflorum* J.J.Sm.

167

Part II: Dendrobium

Dendrobium delicatulum auct. non Kraenzl. Schlechter
Dendrobium junzaingense J.J.Sm.
Dendrobium oreocharis Schltr.
Dendrobium tricostatum Schltr.
Pedilonum begoniicarpum (J.J.Sm.) Rauschert
Pedilonum junzaingense (J.J.Sm.) Rauschert
Pedilonum oreocharis (Schltr.) Rauschert
Pedilonum subacaule (Reinw. ex Lindl.) Rauschert
Pedilonum tricostatum (Schltr.) Rauschert

Distribution: Indonesia, Papua New Guinea, Solomon Islands

Dendrobium subclausum Rolfe
Dendrobium asumburu P.Royen
Dendrobium aurantiflavum P.Royen
Dendrobium mitriferum J.J.Sm.
Dendrobium strictum Ridl.
Dendrobium phlox Schltr.
Pedilonum mitriferum (J.J.Sm.) Rauschert
Pedilonum strictum (Ridl.) Rauschert
Pedilonum subclausum (Rolfe) Rauschert

Distribution: Indonesia, Papua New Guinea

Dendrobium subclausum var. **pandanicola** J.J. Wood

Distribution: Papua New Guinea

Dendrobium subclausum var. **phlox** (Schltr.) J.J. Wood
Pedilonum phlox (Schltr.) Rauschert

Distribution: Papua New Guinea

Dendrobium subclausum var. **speciosum** J.J. Wood
Dendrobium obtusisepalum auct. non J.J.Sm.

Distribution: Papua New Guinea

Dendrobium subclausum Rolfe var. **subclausum**

Distribution: Indonesia, Papua New Guinea

Dendrobium subquadratum J.J.Sm.
Dendrobium sikinii Schltr.
Sayeria subquadrata (J.J.Sm.) Rauschert

Distribution: Indonesia, Papua New Guinea

Dendrobium subulatoides Schltr.

Aporum subulatoides (Schltr.) Raushert

Distribution: Malaysia

Dendrobium subuliferum J.J.Sm.
Dendrobium subuliferum var. *gautierense* J.J.Sm.
Pedilonum subuliferum (J.J.Sm.) Rauschert

Distribution: Indonesia, Papua New Guinea

Dendrobium × suffusum Cady
Dendrobium kingianum var. *suffusum* Cady
Tropilis × suffusa (Cady) Butzin

Distribution: Australia

Dendrobium sulcatum Lindl.
Callista sulcata (Lindl.) Kuntze

Distribution: China, India, Lao People's Democratic Republic, Myanmar, Thailand

Dendrobium sulphureum Schltr.
Pedilonum sulphureum Schltr.

Distribution: Indonesia, Papua New Guinea

Dendrobium sulphureum var. **cellulosum** (J.J.Sm.) T.M.Reeve & P.Woods
Dendrobium cellulosum J.J.Sm.

Distribution: Indonesia, Papua New Guinea

Dendrobium sulphureum var. **rigidifolium** T.M.Reeve & P.Woods

Distribution: Papua New Guinea

Dendrobium sulphureum var. **sulphureum**

Distribution: Indonesia, Papua New Guinea

Dendrobium × superbiens Rchb.f.
Callista fitzgeraldii (F.Muell.) Kuntze
Callista superbiens (Rchb.f.) Kuntze
Dendrobium bigibbum var. *albomarginatum* F.M.Bailey
Dendrobium bigibbum var. *superbiens* (Rchb.f.) F.M.Bailey
Dendrobium brandtiae Kraenzl.
Dendrobium fitzgeraldii F.Muell ex F.M.Bailey
Dendrobium goldiei Rchb.f.
Dendrobium superbiens Fitzg.

169

Part II: Dendrobium

Dendrobium × vinicolor St.Cloud

Distribution: Australia

Dendrobium sutepense Rolfe ex Downie

Distribution: Myanmar, Thailand

Dendrobium swartzii A.D.Hawkes & A.H.Heller
Callista lilacina (Rchb.f.) Kuntze
Dendrobium lilacinum Rchb.f. non Teijsm. & Binn.

Distribution: Malaysia

Dendrobium sylvanum Rchb.f.
Dendrobium kennedyi Schltr.
Dendrobium prionochilum F.Muell. & Kraenzl.
Dendrobium robustum Rolfe
Dendrobium validum Schltr.
Dendrobium warianum Schltr.

Distribution: New Caledonia, Papua New Guinea, Solomon Islands

Dendrobium takahashii Carr

Distribution: Indonesia

Dendrobium tangerinum P.J.Cribb

Distribution: Papua New Guinea

Dendrobium tapiniense T.M.Reeve

Distribution: Papua New Guinea

Dendrobium tarberi M.A.Clem. & D.L.Jones
Dendrobium hillii J.D.Hook.

Distribution: Australia

Dendrobium tattonianum Bateman ex Rchb.f.
Callista tattonianum (Rchb.f.) Kuntze
Dendrobium canaliculatum var. *tattonianum* (Bateman ex Rchb.f.) Rchb.f.

Distribution: Australia

Dendrobium taurinum Lindl.

Callista taurina (Lindl.) Kuntze

Distribution: Philippines (the)

Dendrobium taurulinum J.J.Sm.

Distribution: Indonesia

Dendrobium taveuniense Dauncey & P.J.Cribb
  Dendrobium bullenianum auct. non Rchb.f.
  Dendrobium catillare auct. non Rchb.f.
  Dendrobium erythroxanthum auct. non Rchb.f.
  Dendrobium sertatum auct. non Rolfe

Distribution: Fiji

Dendrobium tenellum (Blume) Lindl.
  Aporum tenellum (Blume) Rauschert
  Callista tenella (Blume) Kuntze
  Onychium tenellum Blume

Distribution: Indonesia

Dendrobium tenue J.J.Sm.
  Aporum tenue (J.J.Sm.) Rauschert

Distribution: Indonesia

Dendrobium teretifolium R.Br.
  Callista teretifolia (R.Br.) Kuntze
  Dockrillia teretifolia (R.Br.) Brieger

Distribution: Australia

Dendrobium terminale Parish & Rchb.f.
  Aporum verlaquii (Costantin) Rauschert
  Callista terminalis (Parish & Rchb.f.) Kuntze
  Dendrobium albayense auct. non Ames
  Dendrobium anceps auct. non Lindl.
  Dendrobium leonis var. *strictum* Finet
  Dendrobium verlaquii Costantin

Distribution: China, India, Malaysia, Myanmar, Thailand, Viet Nam

Dendrobium terrestre J.J.Sm.
  Dendrobium magnificum Schltr.
  Dendrobium terrestre var. *sublobatum* J.J.Sm.
  Sayeria terrestris (J.J.Sm.) Rauschert

Part II: Dendrobium

Distribution: Indonesia, Papua New Guinea

Dendrobium tetrachromum Rchb.f.

Distribution: Indonesia, Malaysia

Dendrobium tetraedre (Blume) Lindl.
Aporum strigosum (Schltr.) Rauschert
Aporum tetraedre (Blume) Brieger
Callista tetraedris (Blume) Kuntze
Dendrobium strigosum
Onychium tetraedre Blume

Distribution: Indonesia

Dendrobium tetragonum Cunn.
Callista tetragona (Cunn.) Kuntze
Tropilis tetragona (Cunn.) Butzin

Distribution: Australia

Dendrobium tetralobum Schltr.
Aporum tetralobum (Schltr.) Rauschert

Distribution: Indonesia

Dendrobium tetrodon Rchb.f. ex Lindl.
Callista tetrodon (Rchb.f. ex Lindl.) Kuntze

Distribution: Indonesia, Thailand

Dendrobium thyrsiflorum Rchb.f. ex Andre
Dendrobium densiflorum auct. non Lindl
Dendrobium densiflorum var. *albolutea* Hook.f.

Distribution: China, India, Lao People's Democratic Republic, Myanmar, Thailand, Viet Nam

Dendrobium thyrsodes Rchb.f.
Dendrobium kuhlii auct. non (Blume) Lindl.
Pedilonum kuhlii auct. non Blume

Distribution: Indonesia

Dendrobium tokai Rchb.f.

Distribution: Fiji, Tonga

172

Dendrobium toressae (F.M.Bailey) Dockrill
Bulbophyllum toressae F.M.Bailey
Dockrillia toressae (F.M.Bailey) Brieger

Distribution: Australia

Dendrobium torricellense Schltr.
Sayeria torricellensis (Schltr.) Rauschert

Distribution: Indonesia, Papua New Guinea

Dendrobium tortile Lindl.
Callista tortilis (Lindl.) Kuntze
Dendrobium dartoisianum De Wild.
Dendrobium haniffii Ridl.
Dendrobium haniffii var. *dartoisianum* (De Wild.)
Dendrobium tortile var. *dartoisianum* (De Wild.) O'Brien
Dendrobium tortile var. *simondii* Gagnep.

Distribution: India, Lao People's Democratic Republic, Malaysia, Myanmar, Nepal, Thailand, Viet Nam

Dendrobium tosaense Makino
Dendrobium pere-fauriei Hayata
Dendrobium tosaense var. *pere-fauriei* (Hayata) Masam.

Distribution: China, China (Taiwan), Japan

Dendrobium tozerensis Lavarack

Distribution: Australia

Dendrobium transparens Wall. ex Lindl.
Callista transparens (Wall.) Kuntze

Distribution: Bhutan, India, Myanmar, Nepal

Dendrobium trachythece Schltr.
Pedilonum trachythece (Schltr.) Rauschert

Distribution: Papua New Guinea

Dendrobium transtilliferum J.J.Sm.

Distribution: Indonesia

Dendrobium trichostomum Rchb.f. ex Oliver
Pedilonum oreogenum (Schltr.) Rauschert

Part II: Dendrobium

Pedilonum trichostomum (Rchb.f. ex Oliver) Rauschert

Distribution: Indonesia

Dendrobium tricuspe (Blume) Lindl.
  Callista tricuspis (Blume) Kuntze
  Onychium tricuspe Blume
  Oporium tricuspis (Blume) Rauschert

Distribution: Indonesia, Malaysia

Dendrobium tridentatum Ames & C.Schweinf.

Distribution: Malaysia

Dendrobium trigonopus Rchb.f.
  Callista trigonopus (Rchb.f.) Kuntze
  Dendrobium velutinum Rolfe

Distribution: China, Lao People's Democratic Republic, Myanmar, Thailand

Dendrobium trilamellatum J.J.Sm.
  Dendrobium johannis var. *semifuscum* Rchb.f.
  Dendrobium semifuscum (Rchb.f.) Lavarack & P.J.Cribb

Distribution: Australia, Indonesia, Papua New Guinea

Dendrobium truncatum Lindl.
  Aporum truncatum (Lindl.) Brieger
  Callista clavipes (Hook.f.) Kuntze
  Callista truncata (Lindl.) Kuntze
  Dendrobium clavipes Hook.f.

Distribution: Indonesia, Malaysia, Thailand, Viet Nam

Dendrobium tubiflorum J.J.Sm.
  Pedilonum tubiflorum (J.J.Sm.) Rauschert

Distribution: Indonesia

Dendrobium uliginosum J.J.Sm.
  Pedilonum uliginosum (J.J.Sm.) Rauschert

Distribution: Indonesia

Dendrobium umbonatum Seidenf.

Distribution: Thailand

Dendrobium uncatum Lindl.
Aporum uncatum (Lindl.) Brieger
Callista uncata (Lindl.) Kuntze

Distribution: Indonesia, Malaysia

Dendrobium uncipes J.J.Sm.
Katherinea uncipes (J.J.Sm.) A.D.Hawkes
Sayeria uncipes (J.J.Sm.) Rauschert

Distribution: Indonesia

Dendrobium undatialatum Schltr.
Dendrobium maboroense Schltr.
Pedilonum maboroense (Schltr.) Rauschert
Pedilonum undatialatum (Schltr.) Rauschert

Distribution: Papua New Guinea, Solomon Islands

Dendrobium unicum Seidenf.
Dendrobium arachnites auct. non Thou. nec Rchb.f.

Distribution: Lao People's Democratic Republic, Thailand, Viet Nam

Dendrobium usterioides Ames

Distribution: Philippines (the)

Dendrobium vagans Schltr.
Dendrobium calamiforme Rolfe
Dendrobium crispatum auct. non Sw. Rchb.f.
Dendrobium seemannii L.O.Williams
Dockrillia vagans (Schltr.) Rauschert

Distribution: Fiji, Solomon Islands

Dendrobium vannouhuysii J.J.Sm.
Pedilonum vannouhuysii (J.J.Sm.) Rauschert

Distribution: Indonesia, Papua New Guinea

Dendrobium ventricosum Kraenzl.
Aporum equitans (Kraenzl.) Brieger
Dendrobium equitans Kraenzl.

Distribution: China (Taiwan), Philippines (the)

Part II: Dendrobium

Dendrobium ventrilabium J.J.Sm.

Distribution: Indonesia

Dendrobium ventripes Carr

Distribution: Malaysia

Dendrobium venustum Teijsm. & Binn.
Callista ciliata (Parish ex Hook.f.) Kuntze
Dendrobium ciliatum Parish ex Hook.f.
Dendrobium ciliatum var. *rupicola* Rchb.f.
Dendrobium ciliferum Bakh.
Dendrobium delacourii auct. non Guillaum.
Dendrobium rupicola Rchb.f.

Distribution: Cambodia, Lao People's Democratic Republic, Myanmar, Thailand, Viet Nam

Dendrobium verruculosum Schltr.
Pedilonum verruculosum (Schltr.) Rauschert

Distribution: Papua New Guinea

Dendrobium vexillarius J.J.Sm.
Dendrobium semeion P.Van Royen
Pedilonum vexillarius (J.J.Sm.) Rauschert

Distribution: Indonesia, Papua New Guinea

Dendrobium vexillarius var. **albiviride** (Van Royen) T.M.Reeve & P.Woods
Dendrobium albiviride Van Royen
Dendrobium albiviride var. *minor* Van Royen

Distribution: Papua New Guinea

Dendrobium vexillarius var. **elworthyi** T.M.Reeve & P.Woods

Distribution: Papua New Guinea

Dendrobium vexillarius var. **microblepharum** (Schltr.) T.M.Reeve & P.Woods
Dendrobium microblepharum Schltr.
Dendrobium vexillarius sensu Cribb, Reeve & Woods
Pedilonum microblepharum (Schltr.) Rauschert

Distribution: Indonesia, Papua New Guinea

Dendrobium vexillarius var. **retroflexum** (J.J.Sm.) T.M.Reeve & P.Woods
Dendrobium caenosicallainum P. Van Royen

Dendrobium pentagonum sensu J.J.Sm.
Dendrobium retroflexum J.J.Sm.
Pedilonum retroflexum (J.J.Sm.) Rauschert

Distribution: Indonesia, Papua New Guinea

Dendrobium vexillarius var. **uncinatum** (Schltr.) T.M.Reeve & P.Woods
Dendrobium bilamellatum R.S.Rogers
Dendrobium brachyphyta Schltr.
Dendrobium tenens J.J.Sm.
Dendrobium trialatum Schltr.
Dendrobium trifolium J.J.Sm.
Dendrobium uncinatum Schltr.
Dendrobium xiphiphorum Van Royen
Pedilonum brachyphyta (Schltr.) Rauschert
Pedilonum trialatum (Schltr.) Rauschert
Pedilonum uncinatum (Schltr.) Rauschert

Distribution: Indonesia, Papua New Guinea

Dendrobium vexillarius var. **vexillarius**

Distribution: Indonesia, Papua New Guinea

Dendrobium victoriae-reginae Loher
Pedilonum victoriae-reginae (Loher) Rauschert

Distribution: Philippines (the)

Dendrobium violaceoflavens J.J.Sm.

Distribution: Indonesia

Dendrobium violaceominiatum Schltr.

Distribution: Micronesia (Federated States of), Solomon Islands

Dendrobium violaceum Kraenzl.
Dendrobium dryadum Schltr.
Dendrobium geminiflorum Schltr.
Dendrobium pityphyllum Schltr.
Dendrobium quinquecostatum Schltr.
Dendrobium tenuicalcar J.J.Sm.
Pedilonum dryadum (Schltr.) Rauschert
Pedilonum geminiflorum (Schltr.) Rauschert
Pedilonum pityphyllum (Schltr.) Rauschert
Pedilonum quinquecostatum (Schltr.) Brieger
Pedilonum tenuicalcar (J.J.Sm.) Rauschert
Pedilonum violaceum (Kraenzl.) Rauschert

Distribution: Indonesia, Papua New Guinea

Dendrobium violaceum subsp. **cyperifolium** (Schltr.) T.M.Reeve & P.Woods
Dendrobium cyperifolium Schltr.
Dendrobium igneoviolaceum J.J.Sm.
Dendrobium scotiiferum J.J.Sm.
Pedilonum cyperifolium (Schltr.) Rauschert
Pedilonum scotiiferum (J.J.Sm.) Rauschert

Distribution: Indonesia, Papua New Guinea

Dendrobium violaceum subsp. **violaceum**

Distribution: Indonesia, Papua New Guinea

Dendrobium violascens J.J.Sm.
Sayeria violascens (J.J.Sm.) Rauschert

Distribution: Indonesia

Dendrobium virgineum Rchb.f.
Callista virginea (Rchb.f.) Kraenzl.
Dendrobium kontumense Gagnep.
Dendrobium sculptum auct. non Rchb.f.

Distribution: Lao People's Democratic Republic, Myanmar, Thailand, Viet Nam

Dendrobium viriditepalum J.J.Sm.

Distribution: Indonesia

Dendrobium × vonpaulsenianum A.Hawkes
Dendrobium × intermedium Schltr.

Distribution: Papua New Guinea

Dendrobium wardianum Warner
Callista wardiana (Warner) Kuntze

Distribution: Bhutan, China, India, Myanmar, Thailand, Viet Nam

Dendrobium wassellii S.T.Blake
Dockrillia wassellii (S.T.Blake) Brieger

Distribution: Australia

Dendrobium wattii (Hook.f.) Rchb.f.

Callista wattii (Hook.f.) Kraenzl.
Dendrobium cariniferum var. *wattii* Hook.f.
Dendrobium evrardii Gagnep.
Dendrobium longicornu auct. non Lindl.

Distribution: India, Lao People's Democratic Republic, Myanmar, Thailand, Viet Nam

Dendrobium wentianum J.J.Sm.
Dendrobium deflexum Ridl.
Dendrobium obtusisepalum J.J.Sm.
Dendrobium vacciniifolium J.J.Sm.
Pedilonum deflexum (Ridl.) Rauschert
Pedilonum obtusisepalum (J.J.Sm.) Rauschert
Pedilonum wentianum (J.J.Sm.) Rauschert

Distribution: Indonesia, Papua New Guinea

Dendrobium wenzelii Ames

Distribution: Philippines (the)

Dendrobium whistleri P.J.Cribb

Distibution: Samoa, Solomon Islands

Dendrobium williamsianum Rchb.f.

Distribution: Papua New Guinea

Dendrobium williamsonii Day & Rchb.f.
Callista lubbersiana (Rchb.f.) Kuntze
Callista williamsonii (Day & Rchb.f.) Kuntze
Dendrobium longicornu auct. non Lindl.
Dendrobium lubbersianum Rchb.f.
Dendrobium wattii auct. non Rchb.f.

Distribution: China, India, Myanmar, Thailand, Viet Nam

Dendrobium wilmsianum Schltr.

Distribution: Myanmar, Thailand

Dendrobium wilsonii Rolfe

Distribution: China

Dendrobium wisselense P.J.Cribb

Part II: Dendrobium

Sayeria wisselensis (P.J. Cribb) Rauschert

Distribution: Indonesia

Dendrobium woluense J.J.Sm.
Pedilonum woluense (J.J.Sm.) Rauschert

Distribution: Indonesia

Dendrobium womersleyi T.M.Reeve

Distribution: Indonesia, Papua New Guinea

Dendrobium womersleyi var. **autophilum** Dauncey

Distribution: Papua New Guinea

Dendrobium womersleyi var. **womersleyi**

Distribution: Indonesia, Papua New Guinea

Dendrobium woodsii P.J.Cribb
Dendrobium fantasticum auct non L.O.Williams
Sayeria woodsii (P.J.Cribb) Rauschert

Distribution: Papua New Guinea

Dendrobium wulaiense Howcroft

Distribution: Papua New Guinea

Dendrobium xanthoacron Schltr.

Distribution: Indonesia, Malaysia

Dendrobium xanthogenium Schltr.
Pedilonum xanthogenium (Schltr.) Rauschert

Distribution: Papua New Guinea

Dendrobium xanthophlebium Lindl.
Callista xanthophlebia (Lindl.) Kuntze
Callista marginata (Lindl.) Kraenzl.
Dendrobium marginatum Bateman

Distribution: Myanmar, Thailand

180

Dendrobium xiphophyllum Schltr.

Distribution: Malaysia

Dendrobium × yengiliense T.M.Reeve

Distribution: Papua New Guinea

Dendrobium ypsilon Seidenf.

Distribution: Thailand

DISA BINOMIALS IN CURRENT USAGE

DISA BINOMES ACTUELLEMENT EN USAGE

DISA BINOMIALES UTILIZADOS NORMALMENTE

Disa aconitoides Sond.

Distribution: Burundi, Congo (the Democratic Republic of the), Ethiopia, Kenya, Malawi, Mozambique, Rwanda, South Africa, Tanzania (the United Republic of), Uganda, Zambia, Zimbabwe

Disa aconitoides subsp. **aconitoides**

Distribution: Burundi, Congo (the Democratic Republic of the), Ethiopia, Kenya, Malawi, Mozambique, Rwanda, South Africa, Tanzania (the United Republic of), Uganda, Zambia, Zimbabwe

Disa aconitoides subsp. **concinna** (N.E.Br.) H.P.Linder
Disa bisetosa Kraenzl.
Disa concinna N.E.Br.
Disa equestris var. *concinna* (N.E.Br) Kraenzl.

Distribution: Malawi, Mozambique, Zambia, Zimbabwe

Disa aconitoides subsp. **goetzeana** (Kraenzl.) H.P.Linder
Disa chiovendaei Schltr.
Disa goetzeana Kraenzl.
Disa vaginata Chiov.

Distribution: Ethiopia, Kenya, Tanzania (the United Republic of), Uganda

Disa aequiloba Summerh.

Distribution: Angola, Congo (the Democratic Republic of the), Tanzania (the United Republic of), Zambia

Disa alticola H.P.Linder

Distribution: South Africa

Disa amoena H.P.Linder

Distribution: South Africa

Disa andringitrana Schltr.

Distribution: Madagascar

Disa aperta N.E.Br.
Disa equestris var. *concinna* (N.E.Br.) Kraenzl.

Distribution: Malawi, Tanzania (the United Republic of), Zambia

Disa arida Vlok

Distribution: South Africa

Disa aristata H.P.Linder

Distribution: South Africa

Disa atricapilla (Harv. ex Lindl.) Bolus
Disa bivalvata var. *atricapilla* (Harv. ex Lindl.) Schltr.
Orthopenthea atricapilla (Harv. ex Lindl.) Rolfe
Penthea atricapilla Harv. ex Lindl.

Distribution: South Africa

Disa aurata (Bolus) Koopowitz & L.T.Parker
Disa tripetaloides subsp. *aurata* (Bolus) H.P.Linder
Disa tripetaloides var. *aurata* Bolus

Distribution: South Africa

Disa basutorum Schltr.
Monadenia basutorum (Schltr.) Rolfe

Distribution: Lesotho, South Africa

Disa begleyi L.Bolus

Distribution: South Africa

Disa bivalvata (L.f.) T.Durand & Schinz
Disa melaleuca (Thunb.) Sw.
Ophrys bivalvata L.f.
Orthopenthea bivalvata (L.f.) Rolfe
Penthea melaleuca (Thunb.) Lindl.
Serapias melaleuca Thunb.

Distribution: South Africa

Disa bodkinii Bolus

Part II: Disa

Orthopenthea bodkinii (Bolus) Rolfe

Distribution: South Africa

Disa borbonica Balf.f. & S.Moore

Distribution: Reunion

Disa brachyceras Lindl.
Disa tenella var. *brachyceras* (Lindl.) Schltr.

Distribution: South Africa

Disa brevipetala H.P.Linder

Distribution: South Africa

Disa buchenaviana Kraenzl.
Disa *rutenbergiana* Kraenzl.
Satyrium calceatum Ridl.

Distribution: Madagascar

Disa caffra Bolus
Disa compta Summerh.
Disa perrieri Schltr.

Distribution: Angola, Congo, (the Democratic Republic of the), Madagascar, Malawi, South Africa, Zambia

Disa cardinalis H.P.Linder

Distribution: South Africa

Disa caulescens Lindl.

Distribution: South Africa

Disa cedarbergensis H.P.Linder

Distribution: South Africa

Disa celata Summerh.

Distribution: Angola, Malawi, Tanzania (the United Republic of), Zambia

Disa cephalotes Rchb.f.

Distribution: Lesotho, South Africa

Disa cephalotes subsp. **cephalotes**

Distribution: Lesotho, South Africa

Disa cephalotes subsp. **frigida** (Schltr.) H.P.Linder
 Disa frigida Schltr.

Distribution: Lesotho, South Africa

Disa chrysostachya Sw.
 Disa gracilis Lindl.

Distribution: Lesotho, South Africa, Swaziland

Disa clavicornis H.P.Linder

Distribution: South Africa

Disa cochlearis Johnson & Liltved

Distribution: South Africa

Disa cooperi Rchb.f.

Distribution: Lesotho, South Africa

Disa cornuta (L.) Sw.
 Disa aemula Bolus
 Disa cornuta var. *aemula* (Bolus) Kraenzl.
 Disa macrantha Sw.
 Orchis cornuta L.
 Satyrium cornutum (L.) Thunb.

Distribution: Lesotho, South Africa, Zimbabwe

Disa crassicornis Lindl.
 Disa jacottetetiae Kraenzl.
 Disa megaceras Hook.f.

Distribution: Lesotho, South Africa

Disa cryptantha Summerh.

Part II: Disa

Distribution: Burundi, Congo, (the Democratic Republic of the), Tanzania (the United Republic of), Zambia

Disa cylindrica (Thunb.) Sw.
Satyrium cylindrica Thunb.

Distribution: South Africa

Disa danielae Greerinck

Distribution: Congo, (the Democratic Republic of the)

Disa dichroa Summerh.
Disa aconitoides var. *dichroa* (Summerh.) Geerinck
Disa concinna var. *dichroa* (Summerh.) Geerinck

Distribution: Congo, (the Democratic Republic of the), Zambia

Disa dracomontana Schelpe ex H.P.Linder

Distribution: South Africa

Disa draconis (L.f.) Sw.
Disa harveiana Lindl.
Orchis draconis L.f.
Satyrium draconis (L.f.) Thunb.

Distribution: South Africa

Disa elegans Sond. ex Rchb.f.
Orthopenthea elegans (Rchb.f.) Rolfe
Penthea elegans Sond.

Distribution: South Africa

Disa eminii Kraenzl.
Disa stolonifera Rendle

Distribution: Burundi, Rwanda, Tanzania (the United Republic of), Uganda, Zambia

Disa englerana Kraenzl.
Disa subscutellifera Kraenzl.

Distribution: Congo, (the Democratic Republic of the), Malawi, Tanzania (the United Republic of), Zambia

Disa equestris Rchb.f.

Disa huillensis Fritsch

Distribution: Angola, Cameroon, Central African Republic, Congo (the Democratic Republic of the), Malawi, Mozambique, Nigeria, Tanzania (the United Republic of), Zambia, Zimbabwe

Disa erubescens Rendle

Distribution: Angola, Burundi, Cameroon, Congo (the Democratic Republic of the), Kenya, Malawi, Mozambique, Nigeria, Rwanda, Sudan, Tanzania (the United Republic of), Uganda, Zambia, Zimbabwe

Disa erubescens subsp. **carsonii** (N.E.Br.) H.P.Linder
Disa carsonii N.E.Br.
Disa erubescens var. *carsonii* (N.E.Br.) H.P.Linder
Disa stolzii Schltr.

Distribution: Burundi, Congo (the Democratic Republic of the), Malawi, Mozambique, Tanzania (the United Republic of), Zambia

Disa erubescens subsp. **erubescens**
Disa erubescens var. *leucantha* Schltr.

Distribution: Angola, Burundi, Cameroon, Congo (the Democratic Republic of the), Kenya, Malawi, Mozambique, Nigeria, Rwanda, Sudan, Tanzania (the United Republic of), Uganda, Zambia, Zimbabwe

Disa esterhuyseniae Schelpe ex H.P.Linder

Distribution: South Africa

Disa extinctoria Rchb.f.

Distribution: South Africa, Swaziland

Disa fasciata Lindl.
Orthopenthea fasciata (Lindl.) Rolfe

Distribution: South Africa

Disa ferruginea (Thunb.) Sw.
Satyrium ferrugineum Thunb.

Distribution: South Africa

Disa filicornis (L.f.) Thunb.
Disa filicornis var. *latipetala* Bolus
Disa patens Sw.
Disa reflexa (Lindl.) Rchb.f.

Part II: Disa

Orchis filicornis L.f.
Penthea filicornis (L.f.) Lindl.
Penthea reflexa Lindl.
Limodorum longicorne Thunb.

Distribution: South Africa

Disa fragrans Schltr.

Distribution: Congo (the Democratic Republic of the), Ethiopia, Kenya, Lesotho, Malawi, Mozambique, Rwanda, South Africa, Sudan, Tanzania (the United Republic of), Uganda, Zimbabwe

Disa fragrans subsp. **deckenii** (Rchb.f.) H.P.Linder
Disa deckenii Rchb.f.
Disa kilimanjarica Rendle

Distribution: Congo (the Democratic Republic of the), Ethiopia, Kenya, Rwanda, Sudan, Tanzania (the United Republic of), Uganda

Disa fragrans subsp. **fragrans**
Disa leucostachys Kraenzl.
Monadenia junodiana Kraenzl.

Distribution: Congo (the Democratic Republic of the), Ethiopia, Kenya, Lesotho, Malawi, Mozambique, Rwanda, South Africa, Sudan, Tanzania (the United Republic of), Uganda, Zimbabwe

Disa galpinii Rolfe
Disa minax Kraenzl.

Distribution: Lesotho, South Africa

Disa gladioliflora Burch. ex Lindl

Distribution: South Africa

Disa gladioliflora subsp. **capricornis** (Rchb.f.) H.P.Linder
Disa capricornis Rchb.f.
Disa gladioliflora Lindl.

Distribution: South Africa

Disa gladioliflora subsp. **gladioliflora**

Distribution: South Africa

Disa glandulosa Burch. ex Lindl.

Distribution: South Africa

Disa hallackii Rolfe
Disa stokoei L.Bolus

Distribution: South Africa

Disa harveiana Johnson & Linder

Distribution: South Africa

Disa harveiana subsp. **harveiana**

Distribution: South Africa

Disa harveiana subsp. **longicalcarata** Johnson & Linder

Distribution: South Africa

Disa hircicornis Rchb.f.
Disa amblyopetala Schltr.
Disa culveri Schltr.
Disa laeta Rchb.f.

Distribution: Angola, Burundi, Cameroon, Congo (the Democratic Republic of the), Kenya, Malawi, Mozambique, Nigeria, Rwanda, South Africa, Sudan, Tanzania (the United Republic of), Zambia, Zimbabwe, Uganda

Disa incarnata Lindl.
Disa fallax Kraenzl.

Distribution: Madagascar

Disa intermedia H.P.Linder

Distribution: Swaziland

Disa introrsa Kurzweil, Liltved & Linder

Distribution: South Africa

Disa karooica Johnson & Linder

Distribution: South Africa

189

Part II: Disa

Disa katangensis De Wild.
Disa erubescens var. *katangensis* (De Wild.) Geerinck
Disa katangensis var. *katangensis* De Wild.

Distribution: Angola, Congo (the Democratic Republic of the), Zambia

Disa lineata Bolus
Disa neglecta Sond.

Distribution: South Africa

Disa longicornu L.f.

Distribution: South Africa

Disa longifolia Lindl.

Distribution: South Africa

Disa maculata L.f.
Schizodium maculatum (L.f.) Lindl.

Distribution: South Africa

Disa maculomarronina McMurtry

Distribution: South Africa

Disa marlothii Bolus

Distribution: South Africa

Disa micropetala Schltr.

Distribution: South Africa

Disa miniata Summerh.

Distribution: Congo (the Democratic Republic of the), Malawi, Mozambique, Tanzania (the United Republic of), Zambia, Zimbabwe

Disa minor (Sond.) Rchb.f.
Orthopenthea minor (Sond.) Rolfe
Penthea minor Sond.

Distribution: South Africa

Disa montana Sond.
Disa poikilantha Kraenzl.
Disa pulchra var. *montana* (Sond.) Schltr.

Distribution: South Africa

Disa neglecta Sond.

Distribution: South Africa

Disa nervosa Lindl.
Disa fanniniae Rolfe

Distribution: Lesotho, South Africa, Swaziland

Disa nigerica Rolfe

Distribution: Cameroon, Congo (the Democratic Republic of the), Nigeria

Disa nivea H.P.Linder

Distribution: South Africa

Disa nyikensis H.P.Linder

Distribution: Malawi, Tanzania (the United Republic of), Zambia

Disa obtusa Lindl.
Disa tabularis Sond.

Distribution: South Africa

Disa obtusa subsp. **hottentotica** H.P.Linder

Distribution: South Africa

Disa obtusa subsp. **obtusa**
Disa tabularis Sond.

Distribution: South Africa

Disa obtusa subsp. **picta** (Sond.) H.P.Linder
Disa pappei Rolfe
Disa picta Sond.

Part II: Disa

Distribution: South Africa

Disa ocellata Bolus
Disa maculata Harv. ex Lindl. non L.f.

Distribution: South Africa

Disa ochrostachya Rchb.f.
Disa adolphi-friderici Kraenzl.
Disa aurantiaca Rchb.f.

Distribution: Angola, Burundi, Cameroon, Congo (the Democratic Republic of the), Kenya, Malawi, Nigeria, Rwanda, Tanzania (the United Republic of), Uganda, Zambia, Zimbabwe

Disa oligantha Rchb.f.
Disa parvilabris Bolus
Orthopenthea triloba (Sond.) Rolfe
Penthea triloba Sond.

Distribution: South Africa

Disa oreophila Bolus

Distribution: South Africa, Lesotho

Disa oreophila subsp. **erecta** H.P.Linder

Distribution: South Africa

Disa oreophila subsp. **oreophila**

Distribution: South Africa, Lesotho

Disa ornithantha Schltr.

Distribution: Angola, Congo (the Democratic Republic of the), Malawi, Mozambique, Tanzania (the United Republic of), Zambia, Zimbabwe

Disa ovalifolia Sond.

Distribution: South Africa

Disa patula Sond.
Disa nervosa Lindl.
Disa stenoglossa Bolus

Distribution: South Africa, Swaziland, Zimbabwe

Disa patula var. **patula**

Distribution: South Africa, Swaziland, Zimbabwe

Disa patula var. **transvaalensis** Summerh.
  Disa gerrardii Rolfe

Distribution: South Africa, Swaziland, Zimbabwe

Disa perplexa H.P.Linder

Distribution: Cameroon, Congo (the Democratic Republic of the), Kenya, Malawi, Tanzania (the United Republic of), Zambia, Zimbabwe

Disa pillansii L.Bolus

Distribution: South Africa

Disa polygonoides Lindl.
  Disa chrysostachya auct. non Sw.
  Disa natalensis Lindl.

Distribution: Mozambique, South Africa, Swaziland, Zimbabwe

Disa porrecta Sw.
  Disa zeyheri Sond.

Distribution: Lesotho, South Africa

Disa pulchella Hochst. ex A.Rich.

Distribution: Ethiopia, Yemen

Disa pulchra Sond.
  Disa kraussii Rolfe

Distribution: South Africa

Disa racemosa L.f.
  Disa racemosa var. *isopetala* Bolus
  Disa secunda (Thunb.) Sw.
  Satyrium secundatum Thunb.

Distribution: South Africa

Disa rhodantha Schltr.

Distribution: South Africa, Zimbabwe

Disa richardiana Lehm. ex Bolus
Disa richardiana Lehm. ex Lindl.
Orthopenthea obtusa (Lindl.) Schelpe
Orthopenthea richardiana (Bolus) Rolfe
Penthea obtusa Lindl.

Distribution: South Africa

Disa robusta N.E.Br.
Disa coccinea Kraenzl.
Disa praestans Kraenzl.

Distribution: Burundi, Congo (the Democratic Republic of the), Malawi, Rwanda, Tanzania (the United Republic of), Zambia

Disa roeperocharoides Kraenzl.
Disa welwitschii auct. non Rchb.f.

Distribution: Congo (the Democratic Republic of the), Zambia

Disa rosea Lindl.
Orthopenthea rosea (Lindl.) Rolfe

Distribution: South Africa

Disa rungweensis Schltr.

Distribution: Malawi, Tanzania (the United Republic of) Zimbabwe

Disa sagittalis (L.f.) Sw.
Disa attenuata Lindl.
Orchis sagittalis L.f.
Satyrium sagittale (L.f.) Thunb.

Distribution: South Africa

Disa salteri G.J.Lewis

Distribution: South Africa

Disa sanguinea Sond.
Disa huttonii Rchb.f.

Distribution: South Africa

Disa sankeyi Rolfe
Disa basutorum Kraenzl.

Distribution: South Africa

Disa satyriopsis Kraenzl.
Disa ochrostachya var. *latipetala* G.Will.

Distribution: Malawi, Tanzania (the United Republic of), Zambia

Disa saxicola Schltr.
Disa uliginosa Kraenzl.

Distribution: Lesotho, Malawi, Mozambique, South Africa, Swaziland, Tanzania (the United Republic of), Zambia, Zimbabwe

Disa schizodioides Sond.
Orthopenthea schizodioides (Sond.) Rolfe

Distribution: South Africa

Disa scullyi Bolus
Disa cooperi var. *scullyi* (Bolus) Schltr.

Distribution: South Africa

Disa scutellifera A.Rich.
Disa schimperi N.E.Br.

Distribution: Ethiopia, Kenya, Nigeria, Sudan, Uganda

Disa similis Summerh.

Distribution: Angola, South Africa, Zambia

Disa stachyoides Rchb.f.
Monadenia leydenburgensis Kraenzl.

Distribution: Lesotho, South Africa, Swaziland

Disa stairsii Kraenzl.
Disa bakeri Rolfe
Disa gregorana Rendle
Disa luxurians Kraenzl.
Disa wissmannii Kraenzl.

Part II: Disa

Distribution: Congo (the Democratic Republic of the), Kenya, Rwanda, Tanzania (the United Republic of), Uganda

Disa stricta Sond.

Distribution: Lesotho, South Africa

Disa subtenuicornis H.P.Linder

Distribution: South Africa

Disa telipogonis Rchb.f.
  Orthopenthea telipogonis (Rchb.f.) Rolfe

Distribution: South Africa

Disa tenella (L.f.) Sw.
  Orchis tenella L.f.
  Orchis tenuifolia Burm.f.
  Satyrium tenellum (L.f.) Thunb.

Distribution: South Africa

Disa tenella subsp. **pusilla** H.P.Linder

Distribution: South Africa

Disa tenella subsp. **tenella**

Distribution: South Africa

Disa tenuicornis Bolus

Distribution: South Africa

Disa tenuifolia Sw.
  Disa lutea H.P. Linder
  Disa patens (L.f.) Thunb.
  Ophrys patens L.f.
  Penthea patens (L.f.) Lindl.
  Serapias patens (L.f.) Thunb.

Distribution: South Africa

Disa tenuis Lindl.
  Amphigena leptostachys (Sond.) Rolfe
  Amphigena tenuis (Lindl.) Rolfe

Disa leptostachys Sond.
Monadenia tenuis (Lindl.) Kraenzl.

Distribution: South Africa

Disa thodei Schltr. ex Kraenzl.

Distribution: South Africa

Disa triloba Lindl.
Disa sagittalis var. *triloba* (Lindl.) Schltr.

Distribution: South Africa

Disa tripetaloides (L.f.) N.E.Br.
Disa excelsa (Thunb.) Sw.
Disa falcata Schltr.
Disa venosa Lindl.
Orchis tripetaloides L.f.
Satyrium excelsum Thunb.

Distribution: South Africa

Disa tysonii Bolus

Distribution: Lesotho, South Africa

Disa ukingensis Schltr.
Disa englerana auct. non Kraenzl.

Distribution: Malawi, Tanzania (the United Republic of), Zambia

Disa uncinata Bolus

Distribution: South Africa

Disa uniflora P.J.Bergius
Disa grandiflora L.f.
Satyrium grandiflora (L.f.) Thunb.

Distribution: South Africa

Disa vaginata Harv. ex Lindl.
Disa modesta Rchb.f.

Distribution: South Africa

Part II: Disa

Disa vasselotii Bolus ex Schltr.

Distribution: South Africa

Disa venosa Sw.
Disa racemosa var. *venosa* (Sw.) Schltr.

Distribution: South Africa

Disa verdickii De Wild.

Distribution: Angola, Congo (the Democratic Republic of the), Zambia

Disa versicolor Rchb.f.
Disa hemispaerophora Rchb.f.
Disa macowanii Rchb.f.

Distribution: Angola, Lesotho, Mozambique, South Africa, Swaziland, Zimbabwe

Disa walleri Rchb.f.
Disa leopoldii Kraenzl.
Disa princeae Kraenzl.
Disa zombaensis Rendle

Distribution: Burundi, Congo (the Democratic Republic of the), Malawi, Tanzania (the United Republic of), Zambia, Zimbabwe

Disa welwitschii Rchb.f.

Distribution: Angola, Burundi, Cameroon, Central African Republic, Congo (the Democratic Republic of the), Côte d'Ivoire, Guinea, Kenya, Liberia, Malawi, Mozambique, Nigeria, South Africa, Tanzania (the United Republic of), Uganda, Zambia, Zimbabwe

Disa welwitschii subsp. **occultans** (Schltr.) H.P.Linder
Disa occultans Schltr.
Disa subaequalis Summerh.
Disa tanganyikensis Summerh.

Distribution: Cameroon, Central African Republic, Congo (the Democratic Republic of the), Côte d'Ivoire, Guinea, Kenya, Liberia, Malawi, Nigeria, Tanzania (the United Republic of), Uganda, Zambia, Zimbabwe

Disa welwitschii subsp. **welwitschii**
Disa breyeri Schltr.
Disa calophylla Kraenzl.
Disa hyacinthina Kraenzl.
Disa ignea Kraenzl.
Disa welwitschii var. *buchneri* Schltr.

Distribution: Angola, Burundi, Cameroon, Central African Republic, Congo (the Democratic Republic of the), Côte d'Ivoire, Guinea, Kenya, Liberia, Malawi, Mozambique, Nigeria, South Africa, Tanzania (the United Republic of), Uganda, Zambia, Zimbabwe

Disa woodii Schltr.

Distribution: South Africa, Swaziland, Zimbabwe

Disa zimbabweensis H.P.Linder
  Disa rungweensis subsp. *rhodesiaca* (Summerh.) Summerh.
  Disa rungweensis var. *rhodesiaca* Summerh.

Distribution: Mozambique, South Africa, Zimbabwe

Disa zombica N.E.Br.
  Disa nyassana Schltr.

Distribution: Burundi, Congo (the Democratic Republic of the), Malawi, Mozambique, Tanzania (the United Republic of), Zambia, Zimbabwe

Disa zuluensis Rolfe

Distribution: South Africa

DRACULA BINOMIALS IN CURRENT USAGE

DRACULA BINOMES ACTUELLEMENT EN USAGE

DRACULA BINOMIALES UTILIZADOS NORMALMENTE

Dracula alcithoe Luer & R.Escobar

Distribution: Colombia, Ecuador

Dracula amaliae Luer & R.Escobar

Distribution: Colombia

Dracula andreettae (Luer) Luer
Masdevallia andreettae Luer

Distribution: Colombia, Ecuador

Dracula anicula Luer & R.Escobar

Distribution: Colombia

Dracula anthracina Luer & R.Escobar

Distribution: Colombia

Dracula aphrodes Luer & R.Escobar

Distribution: Colombia

Dracula bella (Rchb.f.) Luer
Masdevallia bella Rchb.f.

Distribution: Colombia

Dracula bellerophon Luer & R.Escobar

Distribution: Colombia

Dracula benedictii (Rchb.f.) Luer
Dracula hubeinii Luer
Dracula troglodytes (E.Morren) Luer
Masdevallia benedictii Rchb.f.
Masdevallia troglodytes E.Morren

Distribution: Colombia

Dracula berthae Luer & R.Escobar

Distribution: Colombia

Dracula brangeri Luer

Distribution: Colombia

Dracula carcinopsis Luer & R.Escobar

Distribution: Colombia

Dracula carlueri Hermans & P.J.Cribb

Distribution: Costa Rica

Dracula chestertonii (Rchb.f.) Luer
  Masdevallia chestertonii Rchb.f
  Masdevallia macrochila Regel

Distribution: Colombia

Dracula chimaera (Rchb.f.) Luer
  Dracula senilis (Rchb.f.) Luer
  Masdevallia backhousiana Rchb.f.
  Masdevallia chimaera Rchb.f
  Masdevallia chimaera var. *backhousiana* (Rchb.f.) Williams
  Masdevallia chimaera var. *aurantiaca* Hort. ex. Stein.
  Masdevallia chimaera var. *senilis* (Rchb.f.) J.H.Veitch
  Masdevallia senilis Rchb.f.
  Masdevallia wallisii var. *stupenda* Rchb.f.

Distribution: Colombia

Dracula chiroptera Luer & Malo

Distribution: Colombia, Ecuador

Dracula × circe Luer & R.Escobar

Distribution: Colombia

Dracula citrina Luer & R.Escobar

Distribution: Colombia

Part II: Dracula

Dracula cochliops Luer & R.Escobar

Distribution: Colombia

Dracula cordobae Luer

Distribution: Ecuador

Dracula cutis-bufonis Luer & R.Escobar

Distribution: Colombia

Dracula dalessandroi Luer

Distribution: Ecuador

Dracula dalstroemii Luer

Distribution: Ecuador

Dracula decussata Luer & R.Escobar
Dracula niesseniae Ortiz

Distribution: Colombia

Dracula deltoidea (Luer) Luer
Masdevallia deltoidea Luer

Distribution: Ecuador

Dracula diabola Luer & R.Escobar

Distribution: Colombia

Dracula diana Luer & R.Escobar

Distribution: Colombia

Dracula dodsonii (Luer) Luer
Masdevallia dodsonii Luer

Distribution: Ecuador

Dracula erythrochaete (Rchb.f.) Luer
Dracula burbidgeana (Rolfe) Luer

Dracula gaskelliana (Rchb.f.) Luer
Dracula leonum Luer
Masdevallia burbidgeana Rolfe
Masdevallia erythrochaete Rchb.f.
Masdevallia erythrochaete var. *gaskelliana* (Rchb.f.) Woolward
Masdevallia gaskelliana Rchb.f.

Distribution: Costa Rica, Panama

Dracula erythrochaete subsp. **astuta** (Rchb.f.) Luer
Dracula astuta (Rchb.f.) Luer
Dracula gorgo (Rchb.f.) Luer
Masdevallia astuta Rchb.f.
Masdevallia astuta var. *gaskelliana* Stein.
Masdevallia erythrochaete var. *astuta* (Rchb.f.) Woolward
Masdevallia gorgo Rchb.f. ex Kraenzl.

Distribution: Costa Rica

Dracula erythrochaete subsp. **santa-elenae** Hermans

Distribution: Costa Rica

Dracula exasperata Luer & R.Escobar

Distribution: Colombia

Dracula fafnir Luer

Distribution: Ecuador

Dracula felix (Luer) Luer
Masdevallia felix Luer

Distribution: Colombia, Ecuador

Dracula fuligifera Luer

Distribution: Ecuador

Dracula gastrophora Luer & Hirtz

Distribution: Ecuador

Dracula gigas (Luer & Andreetta) Luer
Masdevallia gigas Luer & Andreetta

Distribution: Colombia, Ecuador

Part II: Dracula

Dracula gorgona (Veitch) Luer & R.Escobar
Masdevallia chimaera var. gorgona Hort ex Veitch

Distribution: Colombia

Dracula gorgonella Luer & R.Escobar

Distribution: Colombia

Dracula hawleyi Luer

Distribution: Ecuador

Dracula hirsuta Luer & Andreetta

Distribution: Ecuador

Dracula hirtzii Luer

Distribution: Colombia, Ecuador

Dracula houtteana (Rchb.f.) Luer
Dracula callifera (Schltr.) Luer
Dracula carderiopsis (Kraenzl.) Luer
Dracula mosquerae (F.C.Lehmann & Kraenzl.) Luer
Masdevallia callifera Schltr.
Masdevallia carderi var. mosquerae F.C.Lehm. & Kraenzl.
Masdevallia carderiopsis Kraenzl.
Masdevallia houtteana Rchb.f.
Masdevallia mosquerae F.C.Lehm. & Kraenzl.

Distribution: Colombia

Dracula inaequalis (Rchb.f.) Luer & R.Escobar
Dracula carderi (Rchb.f.) Luer
Masdevallia carderi Rchb.f.
Masdevallia inaequalis Rchb.f.

Distribution: Colombia

Dracula incognita Luer & R.Escobar

Distribution: Colombia

Dracula insolita Luer & R.Escobar

Distribution: Colombia

Dracula janetiae (Luer) Luer
  Masdevallia janetiae Luer

Distribution: Peru

Dracula lafleurii Luer & Dalström

Distribution: Ecuador

Dracula lehmanniana Luer & R.Escobar

Distribution: Colombia

Dracula lemurella Luer & R.Escobar

Distribution: Colombia

Dracula levii Luer

Distribution: Colombia, Ecuador

Dracula ligiae Luer & R.Escobar

Distribution: Colombia

Dracula lindstroemii Luer & Dalström

Distribution: Ecuador

Dracula lotax (Luer) Luer
  Masdevallia bomboiza Fiske
  Masdevallia lotax Luer

Distribution: Ecuador

Dracula mantissa Luer & R.Escobar

Distribution: Colombia, Ecuador

Dracula marsupialis Luer & Hirtz

Distribution: Ecuador

Part II: Dracula

Dracula minax Luer & R.Escobar

Distribution: Colombia

Dracula mopsus (F.C.Lehmann & Kraenzl.) Luer
Masdevallia mopsus F.C.Lehman & Kraenzl.
Masdevallia triceratops Luer

Distribution: Ecuador

Dracula morleyi Luer & Dalström

Distribution: Ecuador

Dracula navarroorum Luer & Hirtz

Distribution: Ecuador

Dracula nosferatu Luer & R.Escobar

Distribution: Colombia

Dracula nycterina (Rchb.f.) Luer
Masdevallia chimaera Linden & Andre
Masdevallia nycterina Rchb.f.

Distribution: Colombia

Dracula octavioi Luer & R.Escobar

Distribution: Colombia

Dracula ophioceps Luer & R.Escobar

Distribution: Colombia

Dracula orientalis Luer & R.Escobar

Distribution: Colombia

Dracula ortiziana Luer & R.Escobar

Distribution: Colombia

Dracula papillosa Luer & Dodson

Distribution: Ecuador

Dracula pholeodytes Luer & R.Escobar

Distribution: Colombia

Dracula × pileus Luer & R.Escobar

Distribution: Colombia

Dracula platycrater (Rchb.f.) Luer
Dracula lowii (Rolfe) Luer
Dracula trinema sensu Woolward & F.Lehm.
Masdevallia lowii Rolfe
Masdevallia platycrater Rchb.f.
Masdevallia trinema sensu Woolward & F.Lehm.

Distribution: Colombia

Dracula polyphemus (Luer) Luer
Masdevallia polyphemus Luer

Distribution: Ecuador

Dracula portillae Luer & Andreetta

Distribution: Ecuador

Dracula posadarum Luer & R.Escobar

Distribution: Colombia

Dracula presbys Luer & R.Escobar

Distribution: Colombia

Dracula psittacina (Rchb.f.) Luer & R.Escobar
Masdevallia psittacina Rchb.f.

Distribution: Colombia

Dracula psyche (Luer) Luer
Masdevallia psyche Luer

Distribution: Ecuador

Part II: Dracula

Dracula pubescens Luer & Dalström

Distribution: Ecuador

Dracula pusilla (Rolfe) Luer
Dracula vagabunda Luer & R.Escobar
Masdevallia johannis Schltr.
Masdevallia pusilla Rolfe

Distribution: Costa Rica, Guatemala, Mexico, Nicaragua, Panama

Dracula radiella Luer
Dracula fuliginosa (Luer) Luer
Masdevallia fuliginosa Luer

Distribution: Ecuador

Dracula radiosa (Rchb.f.) Luer
Dracula medellinensis (Kraenzl.) Luer
Masdevallia medellinensis Kraenzl.
Masdevallia radiosa Rchb.f.

Distribution: Colombia, Ecuador

Dracula × radio-syndactyla Luer

Distribution: Colombia

Dracula rezekiana Luer & Hawley

Distribution: Ecuador

Dracula ripleyana Luer

Distribution: Costa Rica

Dracula robledorum (P.Ortiz) Luer & R.Escobar
Masdevallia chimaera var. *robledorum* P.Ortiz

Distribution: Colombia

Dracula roezlii (Rchb.f.) Luer
Masdevallia chimaera var. *roezlii* (Rchb.f.) J.H.Veitch
Masdevallia chimaera var. *winniana* (Rchb.f.) J.H.Veitch
Masdevallia roezlii Rchb.f.
Masdevallia roezlii var. *rubrum* Williams
Masdevallia winniana Rchb.f.

208

Distribution: Colombia

Dracula sergioi Luer & R.Escobar

Distribution: Colombia

Dracula severa (Rchb.f.) Luer
  Masdevallia chimaera var. *severa* (Rchb.f.) J.H.Veitch
  Masdevallia severa Rchb.f.
  Masdevallia spectrum Rchb.f.

Distribution: Colombia

Dracula sibundoyensis Luer & R.Escobar

Distribution: Colombia, Ecuador

Dracula simia (Luer) Luer

Distribution: Ecuador

Dracula sodiroi (Schltr.) Luer
  Masdevallia sodiroi Schltr.

Distribution: Ecuador

Dracula syndactyla Luer

Distribution: Colombia

Dracula trichroma (Schltr.) Hermans
  Dracula quilichaoensis (F.C.Lehmann & Kraenzl.) Luer
  Masdevallia iricolor Rchb.f.
  Masdevallia quilichaoensis F.C.Lehm. & Kraenzl.
  Masdevallia trichroma Schltr.
  Masdevallia tricolor Rchb.f.
  Dracula iricolor (Rchb.f.) Luer & R.Escobar

Distribution: Colombia, Ecuador

Dracula trinympharum Luer

Distribution: Ecuador

Dracula tubeana (Rchb.f.) Luer
  Dracula tarantula (Luer) Luer
  Masdevallia tarantula Luer

Part II: Dracula

Masdevallia tubeana Rchb.f

Distribution: Ecuador

Dracula ubangina Luer & Andreetta

Distribution: Ecuador

Dracula vampira (Luer) Luer
Masdevallia vampira Luer

Distribution: Ecuador

Dracula velutina (Rchb.f.) Luer
Dracula lactea (Kraenzl.) Luer
Dracula microglochin (Rchb.f.) Luer
Dracula trinema (Rchb.f.) Luer
Masdevallia lactea Kraenzl.
Masdevallia microglochin Rchb.f.
Masdevallia trinema Rchb.f.
Masdevallia velutina Rchb.f.

Distribution: Colombia

Dracula venefica Luer & R.Escobar
Masdevallia venosa Rolfe

Distribution: Colombia

Dracula venosa (Rolfe) Luer

Distribution: Colombia, Ecuador

Dracula verticulosa Luer & R.Escobar

Distribution: Colombia

Dracula vespertilio (Rchb.f.) Luer
Masdevallia vespertilio Rchb.f.

Distribution: Colombia, Costa Rica, Ecuador, Nicaragua

Dracula vinacea Luer & R.Escobar

Distribution: Colombia

Dracula vlad-tepes Luer & R.Escobar

Distribution: Colombia

Dracula wallisii (Rchb.f.) Luer
Masdevallia chimaera var. *wallisii* (Rchb.f.) J.H.Veitch
Masdevallia wallisii Rchb.f.
Masdevallia wallisii var. *discoidea* Rchb.f.

Distribution: Colombia, Ecuador

Dracula woolwardiae (Lehm.) Luer
Masdevallia woolwardiae Lehm.

Distribution: Ecuador

Dracula xenos Luer & R.Escobar

Distribution: Colombia

ENCYCLIA BINOMIALS IN CURRENT USAGE

ENCYCLIA BINOMES ACTUELLEMENT EN USAGE

ENCYCLIA BINOMIALES UTILIZADOS NORMALMENTE

Encyclia abbreviata (Schltr.) Dressler
Epidendrum abbreviatum Schltr.
Epidendrum prorepens Ames

Distribution: Colombia, Costa Rica, Ecuador, Guatemala, Honduras, Mexico, Nicaragua, Panama

Encyclia acuta Schltr
Epidendrum acutum (Schltr.) Hawkes

Distribution: Venezuela, Brazil

Encyclia adenocarpon (La Llave & Lex.) Schltr.
Epidendrum adenocarpon La Llave & Lex.
Epidendrum adenocarpon subsp. *papillosa* E.Aguirre
Epidendrum adenocarpon var. *rosei* Ames, F.T.Hubb. & C.Schweinf.
Epidendrum crispatum Knowles & Westc.
Epidendrum papillosum Bateman

Distribution: Ecuador, El Salvador, Guatemala, Honduras, Mexico, Nicaragua

Encyclia adenocaula (La Llave & Lex.) Schltr.
Encyclia nemoralis (Lindl.) Schltr.
Epidendrum adenocaulum La Llave & Lex.
Epidendrum nemorale Lindl.
Epidendrum verrucosum Lindl.

Distribution: Mexico

Encyclia adenocaula var. **kennedyi** (Fowlie & Withner) Hágsater

Distribution: Mexico

Encyclia advena (Rchb.f.) Porto & Brade
Epidendrum advena Rchb.f.

Distribution: Brazil

Encyclia aemula (Lindl.) Carnevali & I.Ramírez
Epidendrum aemulum Lindl.
Epidendrum aemulum var. *brevistriatum* (Lindl.) Rchb.f.
Epidendrum fragans var. *aemulum* (Lindl.) Barb.Rodr.
Epidendrum fragans var. *brevistriatum* (Rchb.f.) Cogn.
Epidendrum fragans spp. *aemula* (Lindl.) Dressler

Distribution: Peru, Venezuela

Encyclia aenicta Dressler & G.E.Pollard
Epidendrum pollardianum Withner

Distribution: Guatemala, Honduras, Mexico, Nicaragua

Encyclia alagoensis (Pabst) Pabst
Anacheilium alagoense (Pabst) Pabst, Moutinho & Pinto
Epidendrum alagoense (Pabst)

Distribution: Brazil

Encyclia alata (Bateman) Schltr.
Encyclia alata var. *virella* Dressler & Pollard
Epidendrum alatum Bateman
Epidendrum alatum var. *arrogans* Hort. ex Sander
Epidendrum alatum var. *grandiflorum* Regel
Epidendrum alatum var. *longipetalum* (Lindl. & Paxton) Regel
Epidendrum alatum var. *viridiflorum* Regel
Epidendrum calocheilum Hook.
Epidendrum formosum Klotzsch
Epidendrum longipetalum Lindl. & Paxton

Distribution: Colombia, Costa Rica, Guatemala, Honduras, Mexico, Nicaragua

Encyclia albopurpurea (Barb.Rodr.) Porto & Brade
Epidendrum albopurpureum Barb.Rodr.

Distribution: Brazil

Encyclia alboxanthina Fowlie

Distribution: Brazil

Encyclia allemanii (Barb.Rodr.) Pabst
Anacheilium (Barb.Rodr.) Pabst, Moutinho & Pinto
Epidendrum allemanii Barb.Rodr.
Hormidium allemanii (Barb.Rodr.) Brieger

Distribution: Brazil

Encyclia allemanoides (Hoehne) Pabst
Anacheilium allemanoides (Hoehne) Pabst, Moutinho & Pinto
Epidendrum allemanoides Hoehne
Hormidium allemanoides (Hoehne) Brieger

Distribution: Brazil

Part II: Encyclia

Encyclia amanda (Ames) Dressler

Distribution: Nicaragua, Panama

Encyclia ambigua (Lindl.) Schltr.
Encyclia trachychila (Lindl.) Schltr.
Epidendrum ambiguum Lindl.
Epidendrum trachychilum Lindl.

Distribution: El Salvador, Guatemala, Honduras, Mexico, Nicaragua

Encyclia amicta (Linden & Rchb.f.) Schltr
Epidendrum amictum Linden & Rchb.f.

Distribution: Brazil, Venezuela

Encyclia andrichii L.C.Menezes

Distribution: Brazil

Encyclia angustiloba Schltr.

Distribution: Ecuador, Venezuela

Encyclia apuahuensis (Mansf.) Pabst
Epindendrum apuahuense Mansf.

Distribution: Brazil

Encyclia argentinensis (Speg.) Hoehne
Epindendrum argentinense Speg.
Epindendrum saltensis Hoehne

Distribution: Argentina, Brazil

Encyclia arminii (Rchb.f.) Carnevali & Ramírez
Epidendrum arminii Rchb.f.

Distribution: Colombia, Venezuela

Encyclia aspera (Lindl.) Schltr
Epidendrum asperum Lindl.

Distribution: Ecuador

Encyclia asperirachis Garay

Distribution: Colombia

Encyclia asperula Dressler & G.E.Pollard

Distribution: Belize, Guatemala, Mexico

Encyclia atrorubens (Rolfe) Schltr.
Encyclia diota subsp. *atrorubens* (Rolfe) Dressler & G.E.Pollard
Encyclia diota subsp. *diota* (Lindl.) Schltr.
Epidendrum atrorubens Rolfe

Distribution: Colombia, El Salvador, Mexico

Encyclia auyantepuiensis Carnevali & Ramírez

Distribution: Venezuela

Encyclia baculus (Rchb.f.) Dressler & G.E.Pollard
Encyclia pentotis (Rchb.f.) Dressler
Epidendrum acuminatum Sesse & Moc.
Epidendrum baculus Rchb.f.
Epidendrum beyrodtianum Schltr.
Epidendrum confusum Rolfe
Epidendrum fragrans var. *megalanthum* Lindl.
Epidendrum pentotis Rchb.f.

Distribution: Brazil, Colombia, Costa Rica, El Salvador, Guatemala, Honduras, Mexico, Nicaragua

Encyclia bicamerata (Rchb.f.) Dressler & G.E.Pollard
Epidendrum bicameratum Rchb.f.
Epidendrum karwiushii Rchb.f.

Distribution: Mexico

Encyclia boothiana (Lindl.) Dressler
Diacrium bidentatum (Lindl.) Hemsl.
Encyclia boothiana var. *erythronioides* (Small) Luer
Epicladium boothianum (Lindl.) Small
Epicladium boothianum var. *erythronioides* (Small) Acuña
Epidendrum bidentatum Lindl.
Epidendrum boothianum Lindl.
Epidendrum erythronioides Small
Hormidium boothianum (Lindl.) Brieger

Distribution: Bahamas (the), Cuba, Honduras, Mexico, United States of America (the)

Part II: Encyclia

Encyclia boothiana subsp. **boothiana**

Distribution: Bahamas (the), Cuba, Honduras, Mexico, United States of America (the)

Encyclia boothiana subsp. **favoris** (Rchb.f.) Dressler & G.E.Pollard
Encyclia favoris (Rchb.f.) Soto Arenas

Distribution: Mexico

Encyclia brachiata (A.Rich. & H.Galeotti) Dressler & Pollard
Epidendrum brachiatum A.Rich. & H.G.Galeotti

Distribution: Mexico

Encyclia bracteata (Barb.Rodr.) Hoehne
Epidendrum bracteatum Barb.Rodr.
Epidendrum pusillum Rolfe

Distribution: Brazil

Encyclia bractescens (Lindl.) Hoehne
Encyclia acicularis (Bateman) Schltr.
Epidendrum aciculare Bateman ex Lindl.
Epidendrum bractescens Lindl.
Epidendrum esculentum Hort. ex Lindl.
Epidendrum linearifolium Hook.

Distribution: Cuba, Guatemala, Honduras, Mexico

Encyclia bradfordi (Griseb.) Carnevali & Ramírez
Epidendrum bradfordii Griseb.

Distribution: Trinidad and Tobago, Venezuela

Encyclia bragrançae Ruschi

Distribution: Brazil

Encyclia brassavolae (Rchb.f.) Dressler
Epidendrum brassavolae Rchb.f.

Distribution: Costa Rica, El Salvador, Mexico, Panama

Encyclia buchtienii Schltr.

Distribution: Bolivia

Encyclia burle-marxii Pabst

Distribution: Brazil

Encyclia caetensis (Bicalho) Pabst
Anacheilium caetense (Bicalho) Pabst, Moutinho & Pinto
Hormidium caetense Bicalho

Distribution: Brazil

Encyclia calamaria (Lindl.) Pabst
Anacheilium calamarium (Lindl.) Pabst, Moutinho & Pinto
Encyclia organensis (Rolfe) Pabst
Epidendrum calamarium Lindl.
Epidendrum fragrans var. *rivularium* Barb.Rodr.
Epidendrum organense Rolfe
Hormidium calamarium (Lindl.) Brieger

Distribution: Brazil, Suriname, Venezuela

Encyclia campos-portoi Pabst
Anacheilium campos-portoi (Pabst) Pabst, Moutinho & Pinto
Hormidium campos-portoi (Pabst) Brieger

Distribution: Brazil

Encyclia campylostalix (Lindl.) Schltr.
Epidendrum campylostalix Rchb.f.

Distribution: Mexico, Panama

Encyclia candollei (Lindl.) Schltr.
Encyclia flabellata (Lindl.) Thurston
Encyclia laxa Schltr.
Epidendrum asperum auct. non Lindl.
Epidendrum candollei Lindl.
Epidendrum flabellatum Lindl.

Distribution: Costa Rica, Guatemala, Mexico, Panama

Encyclia cardimii Pabst

Distribution: Brazil

Encyclia carpatiana (L.Linden) Fowlie & Duv.
Epidendrum carpatianum L.Linden
Epidendrum godseffianum Rolfe

Distribution: Brazil

Encyclia cepiforme Hooker

Distribution: Guatemala, Mexico

Encyclia ceratistes (Lindl.) Schltr.
Encyclia oncidioides var. *ramonensis* (Rchb.f.) Hoehne
Encyclia peraltensis (Ames) Withner
Encyclia powellii Schltr.
Encyclia ramonensis (Rchb.f.) Schltr.
Epidendrum ceratistes Lindl.
Epidendrum oncidioides var. *ramonense* (Rchb.f.) Ames, F.T.Hubb. & C.Schweinf.
Epidendrum ramonense Rchb.f.

Distribution: Colombia, Costa Rica, Honduras, Mexico, Nicaragua, Panama, Venezuela

Encyclia chacaoensis (Rchb.f.) Dressler & G.E.Pollard
Encyclia ionophlebia (Rchb.f.) Dressler
Encyclia hoffmanii Schltr.
Epidendrum chacaoense Rchb.f.
Epidendrum madrense Schltr.
Epidendrum pachycarpum Schltr.

Distribution: Colombia, Mexico, Nicaragua, Panama, Venezuela

Encyclia chapadensis L.C.Menezes

Distribution: Brazil

Encyclia chiapasensis Withner & Hunt

Distribution: Mexico

Encyclia chimborazoensis (Schltr.) Dressler
Epidendrum chimborazoense Schltr.

Distribution: Colombia, Ecuador, Panama, Venezuela

Encyclia chloroleuca (Hook.) Neum.
Epidendrum chloroleucum W.J.Hook.
Encyclia thienii Dodson

Distribution: Ecuador, Guyana, Peru, Surinam

Encyclia chondylobulbon (A.Rich. & H.Galeotti) Dressler & Pollard
Epidendrum chondylobulbon A.Rich. & H.G.Galeotti

Distribution: El Salvador, Guatemala, Mexico

Encyclia citrina (La Llave & Lex.) Dressler
  Cattleya citrina (La Llave & Lex.) Lindl.
  Cattleya karwinskii Mart.
  Epidendrum citrinum (La Llave & Lex.) Rchb.f.
  Sobralia citrina La Llave & Lex.

Distribution: Mexico

Encyclia cochleata (L.) Lemée
  Anacheilium cochleatum (L.) Hoffmanns.
  Anacheilium cochleatum subsp. *arrogans* (Ames) Small
  Aulizeum cochleatum (L.) Lindl. ex Stein
  Epidendrum cochleatum L.
  Epidendrum cochleatum var. *arrogans* Ames
  Epidendrum cochleatum var. *costaricense* Schltr.
  Epidendrum cochleatum var. *pallidum* Lindl.
  Epidendrum cochleatum var. *triandrumi* Ames
  Epidendrum triandrum (Ames) House
  Hormidium cochleatum (L.) Brieger
  Phaedrosanthus cochleatus (L.) Kuntze

Distribution: Bahamas (the), Colombia, Cuba, El Salvador, Guyana, Mexico, Panama, United States of America (the), Venezuela

Encyclia conchaechila (Barb.Rodr.) Porto & Brade
  Epidendrum conchaechilum Barb.Rodr.

Distribution: Brazil, Venezuela

Encyclia concolor (La Llave & Lex.) Schltr.
  Encyclia amabilis (Linden & Rchb.f.) Schltr.
  Encyclia pruinosa (A.Rich. & H.G.Galeotti) Schltr.
  Epidendrum amabile Linden & Rchb.f.
  Epidendrum concolor La Llave & Lex.
  Epidendrum pruinosum A.Rich. & H.G.Galeotti
  Epidendrum punctulatum Rchb.f.

Distribution: Mexico

Encyclia confusa L.C.Menezes

Distribution: Brazil

Encyclia cordigera (Kunth) Dressler
  Cymbidium cordigerum Kunth
  Encyclia atropurpurea auct. non Schltr.
  Encyclia atropurpurea var. *rosea* (Bateman) Summerh.
  Encyclia atropururea var. *leucantha* Schltr.
  Encyclia atropururea var. *rhodoglossa* Schltr.
  Encyclia cordigera var. *rosea* (Bateman) H.G.Jones

Part II: Encyclia

Encyclia macrochila (Hook.) Neumann
Epidendrum atropurpureum auct. non Willd.
Epidendrum atropurpureum var. *roseum* (Bateman) Rchb.f.
Epidendrum atropurureum var. *laciniatum* Ames, F.T.Hubb. & C.Schweinf.
Epidendrum atropurureum var. *lionetianum* Cogn.
Epidendrum atropurureum var. *longilabre* Cogn.
Epidendrum cordigerum (Kunth) Foldats
Epidendrum longipetalum God.-Leb.
Epidendrum macrochilum Hook.
Epidendrum macrochilum var. *albopurpureum* Morren
Epidendrum macrochilum var. *roseum* Bateman

Distribution: Colombia, Costa Rica, El Salvador, Mexico, Panama, Peru, Venezuela

Encyclia cretacea Dressler & G.E.Pollard

Distribution: Mexico

Encyclia cyanocolumna (Ames, F.T.Hubb. & C.Schweinf.) Dressler
Epidendrum cyanocolumna Ames, F.T.Hubb. & C.Schweinf.

Distribution: Mexico

Encyclia cyperifolia (C.Schweinf.) Carnevali & Ramírez
Bletia ensiformis Ruiz & Pavon
Epidendrum cyperifolium C. Schweinf.
Epidendrum microtos var. *grandiflorum* C.Schweinf.

Distribution: Peru

Encyclia dichroma (Lindl.) Schltr.
Encyclia conspicua (Lem.) Porto & Brade
Encyclia jenischiana (Rchb.f.) Porto & Brade
Epidendrum amabile Hort.
Epidendrum biflorum Barb.Rodr.
Epidendrum conspicuum Lem.
Epidendrum dichromum Lindl.
Epidendrum jenischianum Rchb.f.

Distribution: Brazil

Encyclia dickinsoniana (Withner) Withner
Epidendrum dickinsonianum Withner

Distribution: El Salvador, Guatemala, Honduras, Mexico, Nicaragua

Encyclia diota (Lindl.) Schltr.
Encyclia insidiosa (Rchb.f.) Schltr.
Epidendrum diotum Lindl.
Epidendrum insidiosum Rchb.f.

Distribution: Colombia, El Salvador, Mexico

Encyclia distantiflora (A.Rich. & H.Galeotti) Dressler & Pollard
Epidendrum distantiflorum A.Rich. & H.G.Galeotti

Distribution: Mexico

Encyclia diurna (Jacq.) Schltr.
Cymbidium diurnum (Jacq.) Sw.
Encyclia wageneri (Klotzch) Schltr.
Epidendrum diurna (Jacq.) Britton & Millsp.
Epidendrum diurnum (Jacq.) Cogn.
Epidendrum diurnum (Jacq.) Rchb.f.
Epidendrum ochranthum A.Rich.
Limodorum diurnum Jacq.

Distribution: Venezuela

Encyclia doeringii Hoehne

Distribution: Brazil

Encyclia dutrai Pabst

Distribution: Brazil

Encyclia duveenii Pabst

Distribution: Brazil

Encyclia ensiformis (Ruiz & Pavon) Mansf.
Bletia ensiformis Ruiz & Pavon

Distribution: Brazil

Encyclia euosma Rchb.f.
Epidendrum euosmum Rchb.f.

Distribution: Brazil

Encyclia faresiana (Bicalho) Pabst
Anachelium faresianum (Bicalho) Pabst
Hormidium faresianum Bicalho

Distribution: Brazil

221

Part II: Encyclia

Encyclia fausta (Rchb.f. ex Cogn.) Pabst
Anacheilium faustum (Rchb.f. ex Cogn.) Pabst, Moutinho & Pinto
Epidendrum faustum Rchb.f. ex Cogn.

Distribution: Brazil

Encyclia flabellifera Hoehne & Schltr.

Distribution: Brazil

Encyclia flava (Lindl.) Porto & Brade
Epidendrum flavum Lindl.

Distribution: Brazil

Encyclia fortunae Dressler

Distribution: Panama

Encyclia fowliei Duv.
Encyclia bahiensis L.C.Menezes

Distribution: Brazil

Encyclia fragrans (Sw.) Lemée
Anacheilium fragrans (Sw.) Acuña
Encyclia fragrans subsp. *aemula* (Lindl.) Dressler
Encyclia fragrans var. *brevistriatum* (Rchb.f.) Cogn.
Epidendrum fragrans Sw.
Epidendrum fragrans var. *pachypus* Schltr.
Epidendrum lineatum Salisb.
Epidendrum vaginatum Sesse & Moc.

Distribution: Brazil, Colombia, Guyana, Jamaica, Mexico, Trinidad and Tobago, Venezuela

Encyclia gallopavina (Rchb.f.) Porto & Brade
Epidendrum gallopavinum Rchb.f.
Epidendrum purpurachylum Barb.Rodr.

Distribution: Brazil

Encyclia garciana (Garay & Dunst.) Carnevali & Ramírez
Epidendrum garcianum Garay & Dunst.

Distribution: Venezuela

Encyclia ghiesbreghtiana (A.Rich. & H.G.Galeotti) Dressler
Epidendrum ghiesbreghtianum A.Rich. & H.G.Galeotti

Distribution: Mexico

Encyclia ghillanyi Pabst

Distribution: Brazil

Encyclia glauca (Knowles & Westc.) Dressler & G.E.Pollard
Amblostoma tridactylum var. *mexicanum* Kraenzl.
Encyclia limbata (Lindl.) Dressler
Epidendrum glaucovirens Ames, F.T.Hubb. & C.Schweinf.
Epidendrum glaucum (Knowles & Westc.) Lindl. non Sw.
Epidendrum limbatum Lindl.
Epithecia glauca Knowles & Westc.
Prosthechea glauca Knowles & Westc.

Distribution: El Salvador, Guatemala, Honduras, Mexico

Encyclia glumacea (Lindl.) Pabst
Anacheilium glumaceum (Lindl.) Pabst, Moutinho & Pinto
Epidendrum almasyi Hoehne
Epidendrum glumaceum Lindl.
Hormidium almaysi (Hoehne) Brieger
Hormidium glumaceum (Lindl.) Brieger

Encyclia gonzalezii L.C.Menzies

Distribution: Brazil

Encyclia goyazensis L.C.Menzies

Distribution: Brazil

Encyclia grammatoglossa (Rchb.f.) Dressler
Anacheilium gramatoglossum (Rchb.f.) Pabst, Moutinho & A.V.Pinto
Epidendrum grammatoglossum Rchb.f.
Epidendrum quadidentatum Lehm. & Kraenzl
Hormidium grammatoglossum (Rchb.f.) Brieger

Distribution: Bolivia, Colombia, Ecuador, Peru, Venezuela

Encyclia granitica (Lindl) Schltr.
Epidendrum graniticum Lindl.

Distribution: Brazil, Guyana, Suriname

223

Part II: Encyclia

Encyclia gravida (Lindl.) Schltr.
Encyclia oncidioides var. *gravida* (Lindl.) Hoehne
Epidendrum gravidum Lindl.
Epidendrum oncidioides var. *gravidum* (Lindl.) Ames, F.T.Hubb. & C.Schweinf.

Distribution: Brazil, Mexico, Venezuela

Encyclia guatemalensis (Klotzsch) Withner
Encyclia belizensis (Rchb.f.) Schltr.
Epidendrum belizense Rchb.f.
Epidendrum guatemalense Klotzsch

Distribution: Honduras, Mexico

Encyclia guianensis Carnevali & G Romero

Distribution: Venezuela, Guyana

Encyclia hanburii (Lindl.) Schltr.
Encyclia atropurpureum var. *roseum* Rchb.f.
Epidendrum hanburii Lindl.

Distribution: Mexico

Encyclia hartwegii (Lindl.) R.Vásquez & Dodson
Anacheilium hartwegii (Lindl.) Pabst, Moutinho & Pinto
Encyclia brachychila (Lindl.) Carnevali & Ramírez
Epidendrum brachychilum Lindl.
Epidendrum fuscum Schltr.
Epidendrum hartwegii Lindl.
Epidendrum pachyanthum Schltr.
Hormidium hartwegii (Lindl.) Breiger

Distribution: Bolivia, Colombia, Ecuador, Peru, Venezuela

Encyclia hastata (Lindl.) Dressler & G.E.Pollard
Epidendrum hastatum Lindl.

Distribution: Mexico

Encyclia hoehnei (Hawkes) Pabst
Encyclia squamata (Barb.Rodr.) Porto & Brade
Epidendrum hoehnei Hawkes
Epidendrum squamatum Barb.Rodr. non. Poiret

Distribution: Brazil

Encyclia hollandiae Fowlie

Distribution: Brazil

Encyclia huebneri Schltr.

Distribution: Brazil

Encyclia incumbens (Lindl.) Mabb.
Encyclia aromatica (Bateman) Schltr.
Epidendrum aromaticum Bateman
Epidendrum incumbens Lindl.
Epidendrum primuloides Bateman

Distribution: El Salvador, Guatemala, Mexico

Encyclia ionophlebium (Rchb.f.) Dressler
Epidendrum ionophlebium Rchb.f.

Distribution: Brazil, Colombia, Mexico, Venezuela

Encyclia inversa (Lindl.) Pabst
Anacheilium inversum (Lindl.) Pabst, Moutinho & Pinto
Encyclia bulbosa (Vell.) Pabst
Epidendrum bulbosum Vell. Rchb.f.
Epidendrum fragrans var. *alticallum* Barb.Rodr.
Epidendrum inversum (Lindl.) Pabst
Epidendrum latro Rchb.f. ex Cogn.
Hormidium inversum (Lindl.) Brieger

Distribution: Brazil, Paraguay

Encyclia ionosma (Lindl.) Schltr.
Epidendrum ionosmum Lindl.

Distribution: Brazil

Encyclia ivonae Carnevali & G.A.Romero

Distribution: Brazil, Venezuela

Encyclia jauana Carnevali & Ramírez

Distribution: Venezuela

Encyclia kautskyi Pabst
Anacheilium kautskyi (Pabst) Pabst, Moutinho & Pinto

Distribution: Brazil

Encyclia kennedyi (Fowlie & Withner) Hágsater
Epidendrum kennedyi Fowlie & Withner

Distribution: Mexico

Encyclia kienastii (Rchb.f.) Dressler & G.E.Pollard
Domingoa kienastii (Rchb.f.) Dressler
Epidendrum kienastii Rchb.f.

Distribution: Mexico

Encyclia lambda (Linden & Rchb.f.) Dressler
Epidendrum lambda Linden & Rchb.f.
Epidendrum rueckerae Rchb.f.

Distribution: Colombia

Encyclia lancifolia (Lindl.) Dressler & G.E.Pollard
Epidendrum lancifolium Pav. ex Lindl.
Epidendrum langlassei Schltr.
Epidendrum trulla Rchb.f.

Distribution: Mexico

Encyclia latipetala (C.Schweinf.) Pabst
Epidendrum latipetalum C.Schweinf.

Distribution: Guyana, Brazil, Venezuela

Encyclia laxa Schltr.

Distribution: Brazil

Encyclia leucantha Schltr.
Epidendrum leucanthum (Schltr.) Schnee
Epidendrum leucanthum (Schltr.) C.Schweinf.

Distribution: Colombia, Venezuela

Encyclia lindenii (Lindl.) Carnevali & Ramírez
Epidendrum falax Lindl.
Epidendrum fallax var. *flavescens* (Rchb.f.) Schltr.
Epidendrum lindenii Lindl.

Distribution: Colombia, Venezuela

Encyclia linearifolioides (Kranzl.) Hoehne
Epidendrum bicornuta Brade
Epidendrum linearifolioides Kranzl.

Distribution: Brazil, Paraguay

Encyclia linkiana (Klotzsch) Schltr.
Encyclia pastoris Schltr.
Epidendrum linkianum Klotzsch
Epidendrum pastoris Link & Otto non La Llave & Lex
Epidendrum tripterum Lindl.

Distribution: Mexico

Encyclia livida (Lindl.) Dressler
Anacheilium lividum Lindl
Encyclia deamii (Schltr.) Hoehne
Encyclia tesselata (Lindl.) Schltr.
Epidendrum articulatum Klotzsch
Epidendrum condylochilum E.B.J.Lehm. & Kraenzl.
Epidendrum dasytaenia Schltr.
Epidendrum deamii Schltr.
Epidendrum henrici Schltr.
Epidendrum lividum Lindl.
Epidendrum tessellatum Bateman ex Lindl.
Hormidium lividum (Lindl.) Brieger

Distribution: Colombia, Ecuador, El Salvador, Mexico, Panama, Peru, Venezuela

Encyclia longifolia (Barb.Rodr.) Schltr.
Epidendrum longifolium Barb.Rodr.

Distribution: Brazil

Encyclia lorata Dressler & G.E.Pollard

Distribution: Mexico

Encyclia luteorosea (A.Rich. & H.Galeotti) Dressler & Pollard
Encyclia linearis (Ruiz & Pav.) Dressler
Epidendrum lineare Ruiz & Pav. non Jacq.
Epidendrum luteo-roseum A.Rich. & H.G.Galeotti
Epidendrum seriatum Lindl.

Distribution: Colombia, El Salvador, Honduras, Mexico, Nicaragua, Guyana, Peru, Venezuela

Encyclia lutzenbergerii L.C.Menezes

Distribution: Brazil

Part II: Encyclia

Encyclia maderoi Schltr.

Distribution: Colombia

Encyclia maculosa (Ames, F.T.Hubb. & C.Schweinf.) Hoehne
Encyclia guttata (A.Rich. & H.G.Galeotti) Schltr.
Epidendrum guttatum A.Rich. & H.G.Galeotti
Epidendrum maculosum Ames, F.T.Hubb. & C.Schweinf.

Distribution: Guatemala, Mexico

Encyclia magnispatha (Ames, F.T.Hubb. & C.Schweinf.) Dressler
Epidendrum magnispathum Ames, F.T.Hubb. & C.Schweinf.

Distribution: Mexico

Encyclia mapuerae Huber

Distribution: Brazil

Encyclia mariae (Ames) Hoehne
Epidendrum mariae Ames

Distribution: Mexico

Encyclia megalantha (Barb.Rodr.) Porto & Brade
Epidendrum megalanthum Barb.Rodr.

Distribution: Brazil

Encyclia meliosma (Rchb.f.) Schltr.
Epidendrum meliosmum Rchb.f.

Distribution: Mexico

Encyclia michuacana (La Llave & Lex.) Schltr.
Encyclia icthyphylla (Ames) Hoehne
Encyclia virgata (Lindl.) Schltr.
Epidendrum icthyphyllum Ames
Epidendrum michuacanum La Llave & Lex.
Epidendrum virgatum Lindl.
Epidendrum virgatum var. *arrogans* Rchb.f.

Distribution: El Salvador, Guatemala, Honduras, Mexico

Encyclia microbulbon (Hook.) Schltr.
Encyclia ovulum (Lindl.) Schltr.

Encyclia sisyrinchiifolia (A.Rich. & H.G.Galeotti) Schltr.
Epidendrum microbulbon Hook.
Epidendrum ovulum Lindl.
Epidendrum sisyrinchiifolium A.Rich. & H.G.Galeotti

Distribution: Mexico

Encyclia microtos (Rchb.f.) Hoehne
Epidendrum microtos Rchb.f.

Distribution: Ecuador, Peru

Encyclia microxanthina Fowlie

Distribution: Brazil

Encyclia moojenii (Pabst) Pabst
Anacheilium moojenii (Pabst) Pabst, Moutinho & Pinto
Epidendrum moojenii Pabst

Distribution: Brazil

Encyclia mooreana (Rolfe) Schltr.
Encyclia brenesii Schltr.
Encyclia tonduziana Schltr.
Epidendrum mooreanum Rolfe
Epidendrum oncidioides var. *mooreanum* (Rolfe) Ames, F.T.Hubb & C.Schweinf.

Distribution: Costa Rica, El Salvador, Nicaragua, Panama

Encyclia narajapatensis Dodson

Distribution: Ecuador

Encyclia nematocaulon (A.Rich.) Acuña
Encyclia purpusii Schltr.
Encyclia xipheres (Rchb.f.) Schltr.
Epidendrum nematocaulon A.Rich.
Epidendrum xipheres Rchb.f.

Distribution: Cuba, El Salvador, Guatemala, Honduras, Mexico

Encyclia neurosa (Ames) Dressler & Pollard
Epidendrum neurosum Ames

Distribution: Costa Rica, Guatemala, Honduras, Mexico

Encyclia obpyribulbon Hágsater

Distribution: Mexico

Encyclia ochracea (Lindl.) Dressler
Epidendrum ochraceum Lindl.
Epidendrum parviflorum Sesse & Moc.
Epidendrum triste A.Rich. & H.G.Galeotti

Distribution: Costa Rica, El Salvador, Guyana, Honduras, Mexico, Nicaragua

Encyclia oestlundii (Ames, F.T.Hubb.& Schweinf.) Hágsater & Sterm
Epidendrum oestlundii Ames, F.T.Hubb. & C.Schweinf.

Distribution: Mexico

Encyclia oncidioides (Lindl.) Schltr.
Epidendrum oncidioides Lindl.

Distribution: Brazil

Encyclia osmantha (Barb.Rodr.) Schltr.
Epidendrum osmanthum Barb.Rodr.

Distribution: Brazil

Encyclia oxiphylla Schltr.
Epidendrum oxiphyllum (Schltr.) Hawkes

Distribution: Brazil

Encyclia pachyantha (Lindl.) Hoehne
Epidendrum pachyanthum

Distribution: Brazil

Encyclia panthera (Rchb.f.) Schltr.
Epidendrum panthera Rchb.f.
Epidendrum papyriferum Schltr.

Distribution: Guatemala, Mexico

Encyclia papilio (Vell.) Pabst
Anacheilium papilio (Vell.) Pabst, Moutinho & Pinto
Epidendrum papilio Vell.
Hormidium papilio (Vell.) Brieger

Distribution: Brazil

Encyclia papillosa (Bateman) Aguirre-Olav.

Distribution: Guatemala, Honduras, Mexico

Encyclia parviflora Withner
Encyclia alata subsp. *parvifera* (Regel) Dressler & Pollard
Encyclia alata var. *parviflora* Regel

Distribution: Mexico

Encyclia patens Hook.
Encyclia odoratissima (Lindl.)
Encyclia serroniana (Barb.Rodr.) Hoehne
Epidendrum glutinosum C.Schweinf.
Epidendrum serronianum Barb.Rodr.

Distribution: Brazil

Encyclia pauciflora (Barb.Rodr.) Porto & Brade
Epidendrum pauciflorum Barb.Rodr.

Distribution: Brazil

Encyclia peraltense (Ames) Withner
Epidendrum peraltense Ames

Distribution: Brazil, Mexico, Venezuela

Encylia pflanzii Schltr.
Epidendrum pflanzii (Schltr.) Hawkes

Distribution: Bolivia

Encyclia pedra-azulensis L.C.Menezes

Distribution: Brazil

Encyclia perplexa (Ames, F.T.Hubb. & C.Schweinf.) Withner
Encyclia × perplexa (Ames, F.T.Hubb. & C.Schweinf.) Dressler, Pollard
Epidendrum oncidioides var. *perplexum* Ames, F.T.Hubb. & C.Schweinf.

Distribution: Mexico

Encyclia picta (Lindl.) Hoehne
Epidendrum pictum Lindl.

Part II: Encyclia

Distribution: Guyana

Encyclia pipio (Rchb.f.) Pabst
Epidendrum pipio Rchb.f.

Distribution: Brazil

Encyclia pollardiana (Withner) Dressler & G.E.Pollard

Distribution: Mexico

Encyclia polybulbon (Sw.) Dressler
Bulbophyllum occidentale Spreng.
Dinema polybulbon (Sw.) Lindl.
Epidendrum polybulbon Sw.
Epidendrum polybulbon var. *luteo-album* E.Miege

Distribution: Cuba, El Salvador, Guatemala, Honduras, Jamaica, Mexico, Nicaragua

Encyclia porrecta Adams & P.J.Cribb

Distribution: Belize

Encyclia pringlei (Rolfe ex Ames) Schltr.
Epidendrum pringlei Rolfe ex Ames

Distribution: Mexico

Encyclia prismatocarpa (Rchb.f.) Dressler
Epidendrum ionocentrum Rchb.f.
Epidendrum prismatocarpum Rchb.f.
Hormidium prismatocarpum (Rchb.f.) Brieger et al.

Distribution: Costa Rica, Panama

Encyclia pseudopygmaea (Finet) Dressler & G.E.Pollard
Hormidium pseudopygmaeum Finet

Distribution: El Salvador, Guatemala, Mexico, Panama

Encyclia pterocarpa (Lindl.) Dressler
Epidendrum cinnamomeum A.Rich. & H.G.Galeotti
Epidendrum pterocarpum Lindl.

Distribution: Mexico

Encyclia pulcherrima Dodson & Bennett

Distribution: Ecuador, Peru

Encyclia punctifera (Rchb.f.) Pabst
  Epidendrum punctiferum Rchb.f.
  Epidendrum tripunctatum Lindl.

Distribution: Brazil

Encyclia pygmaea (Hook.) Dressler
  Aulizeum pygmaeum Hook.
  Coelogyne triptera Brongn.
  Encyclia triptera (Brongn.) Dressler & G.E.Pollard
  Epidendrum caespitosum Poepp. & Endl.
  Epidendrum monanthum Steud.
  Epidendrum pygmaeum Hook.
  Epidendrum tripterum Lindl.
  Epidendrum uniflorum Lindl.
  Hormidium humile (Cogn.) Schltr.
  Hormidium pygmaeum (Hook.) Benth. & Hook.f.
  Hormidium pygmaeum (Hook.) Brieger
  Hormidium tripterum (Brongn.) Cogn.
  Hormidium uniflorum (Lindl.) Heynh.
  Microstylis humilis Cogn.

Distribution: Brazil, Mexico, Nicaragua, Panama, United States of America (the)

Encyclia radiata (Lindl.) Dressler
  Anacheilium radiatum Lindl.
  Epidendrum marginatum Link, Klotzsch & Otto
  Epidendrum radiatum Lindl.

Distribution: El Salvador, Guatemala, Honduras, Mexico, Venezuela

Encyclia randii (Barb.Rodr.) Porto & Brade
  Epidendrum randii Barb.Rodr.

Distribution: Brazil

Encyclia recurvata Schltr.
  Epidendrum halatum Garay & Dunst

Distribution: Venezuela

Encyclia regnelliana (Hoehne & Schltr.) Pabst
  Epidendrum regnellianum Hoehne & Schltr.

Distribution: Brazil

233

Part II: Encyclia

Encyclia rhombilabia S.Rosillo

Distribution: Mexico

Encyclia rhynchophora (A.Rich. & H.G.Galeotti) Dressler
Epidendrum rhynchophorum A.Rich. & H.G.Galeotti

Distribution: El Salvador, Guatemala, Honduras, Mexico, Nicaragua

Encyclia rufa (Lindl.) Britton & Millsp.
Encyclia bahamensis (Grisebach. Britton & Millsp.)
Epidendrum bahamense Grisebach.
Epidendrum rufum Lindl.
Epidendrum primullinum Bateman ex Lindl.

Distribution: Bahamas (the), Turks and Caicos

Encyclia sceptra (Lindl.) Carnevali & Ramírez
Epidendrum sceptrum Lindl.

Distribution: Colombia, Venezuela

Encyclia schmidtii L.C.Menezes

Distribution: Brazil

Encyclia sclerocladia (Lindl.) ex Rchb.f.) Hoehne
Epidendrum sclerocladium Lindl.

Distribution: Peru, Venezuela

Encyclia seidelii Pabst

Distribution: Brazil

Encyclia selligera (Bateman ex Lindl.) Schltr.
Epidendrum selligerum Bateman ex Lindl.
Epidendrum violodora Gal. ex Lindl.

Distribution: Bahamas (the), El Salvador, Guatemala, Honduras, Mexico, Nicaragua

Encyclia semiaperta Hágsater
Encyclia tripterum Lindl.

Distribution: Mexico

Encyclia sessiflora (Edwall) Pabst

Epidendrum sessiflorum Edwall

Distribution: Brazil

Encyclia sima Dressler
  Epidendrum simum (Dressler) P.Taylor

Distribution: Panama

Encyclia spatella (Rchb.f.) Schltr.
  Epidendrum spatella Rchb.f.

Distribution: Mexico

Encyclia spiritusanctensis L.C.Menezes
  Encyclia megalantha var. *spiritusantensis* L.C.Menezes

Distribution: Brazil

Encyclia spondiadum (Rchb.f.) Dressler
  Epidendrum platycardium Schltr.
  Epidendrum spondiadum Rchb.f.

Distribution: Costa Rica, Panama

Encyclia steinbachii Schltr.

Distribution: Bolivia

Encyclia stellata (Lindl.) Schltr.
  Encyclia alanjensis (Ames) Carnevali & G.Romero
  Encyclia hunteriana Schltr.
  Epidendrum alanjense Ames
  Epidendrum stellatum Lindl.

Distribution: Costa Rica, Nicaragua, Panama

Encyclia suaveolens Dressler

Distribution: Mexico

Encyclia subulatifolia (A.Rich. & H.G.Galeotti) Dressler
  Epidendrum subulatifolium A.Rich. & H.G.Galeotti

Distribution: Mexico

Encyclia suzanensis (Hoehne) Pabst

Anacheilium suzanense (Hoehne) Pabst, Moutinho & Pabst
Epidendrum suzanense Hoehne

Distribution: Brazil

Encyclia tampensis (Lindl.) Small
Epidendrum tampense Lindl.

Distribution: Bahamas (the), United States of America (the)

Encyclia tarumana Schltr.
Epidendrum tarumanum (Schltr.) Hawkes

Distribution: Brazil

Encyclia tigrina (Lindl. ex Linden) Carnevali & Ramírez
Encyclia pamplonense (Rchb.f.) Carnevali
Epidendrum pamplonense Rchb.f.
Epidendrum tigrinum Lindl. ex Linden

Distribution: Brazil, Colombia, Ecuador, Guyana, Venezuela

Encyclia tenuissima (Ames, F.T.Hubb. & C.Schweinf.) Dressler
Epidendrum tenuissimum Ames, F.T.Hubb. & C.Schweinf.

Distribution: Mexico

Encyclia trachycarpa (Lindl. ex Benth) Schltr.
Encyclia adenocarpon subsp. *trachycarpa* (Lindl. ex Benth) Dressler & G.E.Pollard
Epidendrum trachycarpum Lindl. ex Benth.

Distribution: El Salvador, Guyana, Mexico

Encyclia triangulifera (Rchb.f.) Acuña.
Encyclia bipapularis (Rchb.f.) Acuña.
Encyclia moebusii Dietrich.
Epidendrum bipapulare Rchb.f.
Epidendrum tampense var. *amesianum* Correll.
Epidendrum trianguliferum Rchb.f.

Distribution: Cuba

Encyclia tripartita (Vell.) Hoehne
Epidendrum tripartitum Vell.

Distribution: Brazil

Encyclia tripunctata (Lindl.) Dressler
Encyclia diguetii (Ames) Hoehne

Epidendrum diguetii Ames
Epidendrum micropus Rchb.f.
Epidendrum tripunctatum Lindl.

Distribution: Mexico

Encyclia tuerckheimii Schltr.
Epidendrum tuerckheimii (Schltr.) Ames, F.T.Hubb. & C.Schweinf.

Distribution: Guatemala, Mexico

Encyclia unaensis Fowlie

Distribution: Brazil

Encyclia vagans (Ames) Dressler & G.E.Pollard
Epidendrum vagans Ames

Distribution: Costa Rica, El Salvador, Guatemala, Honduras, Mexico, Nicaragua

Encyclia varicosa (Bateman ex Lindl.) Schltr.
Encyclia chiriquensis (Rchb.f.) Schltr.
Encyclia playmatoglossa (Rchb.f.) Schltr.
Epidendrum chiriquense Rchb.f.
Epidendrum phymatoglossum Rchb.f.
Epidendrum quadratum Klotzsch
Epidendrum varicosum Bateman ex Lindl.

Distribution: Costa Rica, El Salvador, Guyana, Mexico, Panama

Encyclia varicosa subsp. **leiobulbon** (Hook.) Dressler & G.E.Pollard
Epidendrum leiobulbon Hook.
Epidendrum ramírezzi Gajon Sanchez

Distribution: Mexico

Encyclia varicosa subsp. **varicosa**
Encyclia chiriquensis (Rchb.f.) Schltr.
Encyclia playmatoglossa (Rchb.f.) Schltr.
Epidendrum chiriquense Rchb.f.
Epidendrum phymatoglossum Rchb.f.
Epidendrum quadratum Klotzsch
Epidendrum varicosum Bateman ex Lindl.

Distribution: Costa Rica, El Salvador, Guyana, Mexico, Panama

Encyclia vellozoana Pabst
Epidendrum ensiforme Vell.
Encyclia ensiformis (Vell.)

Part II: Encyclia

Distribution: Brazil

Encyclia venezuelana (Schltr.) Schltr.
Epidendrum venezuelanum Schltr.

Distribution: Venezuela

Encyclia venosa (Lindl.) Schltr.
Encyclia pastons (La Llave & Lex.) Schltr.
Encyclia wendlandiana (Kraenzl.) Schltr.
Epidendrum ensicaulon A.Rich. & H.G.Galeotti
Epidendrum pastons La Llave & Lex.
Epidendrum venosum Lindl.
Epidendrum wendlandianum Kraenzl.

Distribution: Mexico

Encyclia vespa (Vell.) Dressler
Anaceilium vespa Vell.
Aulizeum variegatum (Vell.) Pabst, Moutinho & Pinto
Epidendrum baculibulbum Schltr.
Epidendrum christi Rchb.f.
Epidendrum coriaceum Focke
Epidendrum crassilabium Poeppig & Endl.
Epidendrum feddeanum Kraenzl.
Epidendrum leopardinum Rchb.f.
Epidendrum longipes Rhcb.f.
Epidendrum pachysepalum Kl.
Epidendrum rhabdobulbon Schltr.
Epidendrum rhopalobulbon Schltr.
Epidendrum saccharatum Kraenzl.
Epidendrum variegatum Hook.
Epidendrum vespa Vell.

Distribution: Bolivia, Brazil, Colombia, Costa Rica, Ecuador, Guyana, Jamaica, Nicaragua, Panama, Peru, Venezuela

Encyclia virens (Lindl.) Schltr.
Epidendrum virens Lindl.

Distribution: Belize, Guatemala, Mexico, Nicaragua

Encyclia viridiflava L.C.Menezes

Distribution: Brazil

Encyclia viridiflora Hook.
Encyclia multiflora Rchb.f. non Hook.
Epidendrum viridiflorum (Hook.) Lindl.

238

Distribution: Brazil

Encyclia vitellina (Lindl.) Dressler
  Epidendrum vitellinum Lindl.

Distribution: El Salvador, Guatemala, Mexico

Encyclia xerophytica Pabst

Distribution: Brazil

Encyclia xipheroides (Kraenzl.) Porto & Brade

Distribution: Brazil

Encyclia xuxiana Fowlie & Duv.

Distribution: Brazil

Encyclia widgrenii (Lindl.) Pabst
  Anacheilium widgrenii (Lindl.) Pabst, Moutinho & Pinto
  Hormidium widgrenii (Lindl.) Brieger
  Epidendrum widgrenii Lindl.

Distribution: Brazil

Encyclia yauperyensis (Barb.Rodr.) Porto & Brade
  Epidendrum yauperyense Barb.Rodr.

Distribution: Bolivia, Brazil

PART III: COUNTRY CHECKLIST
For the genera:

Cymbidium, Dendrobium (selected sections only)*, Disa, Dracula* and *Encyclia*

Troisième partie: LISTE PAR PAYS
Pour les genre:

Cymbidium, Dendrobium (certaines sections sélectionnées)*, Disa, Dracula* et *Encyclia*

Parte III: LISTA POR PAISES
Para el genero:

Cymbidium, Dendrobium (únicamente las secciones seleccionadas)*, Disa, Dracula* y *Encyclia*

PART III: COUNTRY CHECKLIST FOR THE GENERA:
Cymbidium, Dendrobium (selected sections only), *Disa, Dracula* and *Encyclia*

Troisième partie: LISTE PAR PAYS POUR LES GENRE:
Cymbidium, Dendrobium (certaines sections sélectionnées), *Disa, Dracula* et *Encyclia*

Parte III: LISTA POR PAISES PARA EL GENERO:
Cymbidium, Dendrobium (únicamente las secciones seleccionadas), *Disa, Dracula* y *Encyclia*

ANGOLA / ANGOLA (L') / ANGOLA

Disa aequiloba Summerh.
Disa caffra Bolus
Disa celata Summerh.
Disa equestris Rchb.f.
Disa erubescens Rendle
Disa erubescens subsp. **erubescens**
Disa hircicornis Rchb.f.
Disa katangensis De Wild.
Disa ochrostachya Rchb.f.
Disa ornithantha Schltr.
Disa similis Summerh.
Disa verdickii De Wild.
Disa versicolor Rchb.f.
Disa welwitschii Rchb.f.
Disa welwitschii subsp. **welwitschii**

ARGENTINA / ARGENTINE (L') / ARGENTINA (LA)

Encyclia argentinensis (Speg.) Hoehne

AUSTRALIA / AUSTRALIE (L') / AUSTRALIA

Cymbidium canaliculatum R.Br.
Cymbidium madidum Lindl.
Cymbidium suave R.Br.
Dendrobium adae F.M.Bailey
Dendrobium aemulum R.Br.
Dendrobium affine (Decne) Steud.
Dendrobium antennatum Lindl.
Dendrobium bifalce Lindl.
Dendrobium bigibbum Lindl.
Dendrobium bowmanii Benth.
Dendrobium brachypus (Endl.) Rchb.f.
Dendrobium cacatua M.A.Clem. & D.L.Jones
Dendrobium calamiforme Lodd.
Dendrobium canaliculatum R.Br.
Dendrobium canaliculatum var. **canaliculatum**
Dendrobium canaliculatum var. **pallidum** Dockrill
Dendrobium capitisyork M.A.Clem. & D.L.Jones

Dendrobium carrii Rupp & C.T.White
Dendrobium carronii Lavarack & P.J.Cribb
Dendrobium comptonii Rendle
Dendrobium crumenatum Sw.
Dendrobium cucumerinum Macleay ex Lindl.
Dendrobium curvicaule (F.M.Bailey) M.A.Clem. & D.L.Jones
Dendrobium × delicatum (F.M.Bailey) F.M.Bailey
Dendrobium discolor var. broomfieldii (Fitzg.) M.A.Clem. & D.L.Jones
Dendrobium discolor var. discolor
Dendrobium discolor var. fimbrilabium (Rchb.f.) Dockrill
Dendrobium discolor var. fuscum (Fitzg.) Dockrill
Dendrobium discolor Lindl.
Dendrobium dolichophyllum D.L.Jones & M.A.Clem.
Dendrobium fairfaxii F.Muell. & Fitzg.
Dendrobium falcorostrum Fitzg.
Dendrobium fellowsii F.Muell.
Dendrobium fleckeri Rupp & C.T.White
Dendrobium × foederatum St.Cloud
Dendrobium foelschei F.Muell.
Dendrobium gracilicaule F.Muell.
Dendrobium × gracillimum (Rupp) Leaney
Dendrobium × grimesii C.T.White & Summerh.
Dendrobium johannis Rchb.f.
Dendrobium jonesii Rendle
Dendrobium jonesii subsp. bancroftianium (Rchb.f.) M.A.Clem. & D.L.Jones
Dendrobium jonesii subsp. blackburnii (Nicholls) M.A.Clem. & D.L.Jones
Dendrobium jonesii subsp. jonesii
Dendrobium × kestevenii Rupp
Dendrobium kingianum Bidwill ex Lindl.
Dendrobium × lavarackianum M.A.Clem.
Dendrobium lichenastrum (F.Muell.) Kraenzl.
Dendrobium linguiforme Sw.
Dendrobium lithocola D.L.Jones & M.A.Clem.
Dendrobium litorale Schltr.
Dendrobium macropus (Endl.) Rchb.f. ex Lindl.
Dendrobium melaleucaphilum M.A.Clem. & D.L.Jones
Dendrobium mirbelianum Gaudich.
Dendrobium monophyllum F.Muell.
Dendrobium moorei F.Muell.
Dendrobium mortii F.Muell.
Dendrobium nindii W.Hill
Dendrobium nugentii (F.M.Bailey) D.Jones & M.Clements
Dendrobium pedunculatum (Clemesha) D.L.Jones & M.A.Clem.
Dendrobium phalaenopsis Fitzg.
Dendrobium prenticei (F.Muell.) Nicholls
Dendrobium pugioniforme A.Cunn.
Dendrobium racemosum (Nicholls) Clemesha & Dockrill
Dendrobium rex M.A.Clem. & D.L.Jones
Dendrobium rigidum R.Br.
Dendrobium × ruppiosum Clemesha
Dendrobium schneiderae F.M.Bailey

Dendrobium schneiderae var. **major** Rupp
Dendrobium schneiderae var. **schneiderae**
Dendrobium **schoeninum** Lindl.
Dendrobium **smillieae** F.Muell.
Dendrobium **speciosum** Sm.
Dendrobium **striolatum** Rchb.f.
Dendrobium **stuartii** F.M.Bailey
Dendrobium × **suffusum** Cady
Dendrobium × **superbiens** Rchb.f.
Dendrobium **tarberi** M.A.Clem. & D.L.Jones
Dendrobium **tattonianum** Bateman ex Rchb.f.
Dendrobium **teretifolium** R.Br.
Dendrobium **tetragonum** Cunn.
Dendrobium **toressae** (F.M.Bailey) Dockrill
Dendrobium **tozerensis** Lavarack
Dendrobium **trilamellatum** J.J.Sm.
Dendrobium **wassellii** S.T.Blake

BAHAMAS (THE) / BAHAMAS (LES) / BAHAMAS (LAS)

Encyclia **boothiana** (Lindl.) Dressler
Encyclia **boothiana** subsp. **boothiana**
Encyclia **cochleata** (L.) Lemée
Encyclia **rufa** (Lindl.) Britton & Millsp.
Encyclia **selligera** (Lindl.) Schltr.
Encyclia **tampensis** (Lindl.) Small

BANGLADESH / BANGLADESH (LE) / BANGLADESH

Cymbidium **aloifolium** (L.) Sw.
Dendrobium **crumenatum** Sw.
Dendrobium **longicornu** Wall. ex Lindl.

BELIZE / BELIZE (LE) / BELICE

Encyclia **asperula** Dressler & G.E.Pollard
Encyclia **porrecta** Adams & P.J.Cribb
Encyclia **virens** (Lindl.) Schltr.

BHUTAN / BHOUTAN (LE) / BHUTÁN

Cymbidium **bicolor** Lindl.
Cymbidium **bicolor** subsp. **obtusum** Du Puy & P.J.Cribb
Cymbidium **cyperifolium** Wall. ex Lindl.
Cymbidium **cyperifolium** subsp. **cyperifolium**
Cymbidium **devonianum** Paxton
Cymbidium **elegans** Lindl.
Cymbidium **hookerianum** Rchb.f.
Cymbidium **iridioides** D.Don
Cymbidium **lancifolium** Hook.
Cymbidium **munronianum** King & Pantl.

Dendrobium aduncum Wall. ex Lindl.
Dendrobium amoenum Lindl.
Dendrobium anceps Sw.
Dendrobium aphyllum (Roxb.) Fischer
Dendrobium bicameratum Lindl.
Dendrobium chrysanthum Wall.
Dendrobium chryseum Rolfe
Dendrobium chrysotoxum Lindl.
Dendrobium crepidatum Lindl. & Paxton
Dendrobium cumulatum Lindl.
Dendrobium densiflorum Wall. ex Lindl.
Dendrobium devonianum Paxton
Dendrobium eriiflorum Griff.
Dendrobium falconeri Hook.
Dendrobium farmeri Paxton
Dendrobium fimbriatum Hook.
Dendrobium formosum Roxb. ex Lindl.
Dendrobium gibsonii Lindl.
Dendrobium heterocarpum Lindl.
Dendrobium jenkinsii Wall. ex Lindl.
Dendrobium lindleyi Steud.
Dendrobium longicornu Wall. ex Lindl.
Dendrobium moschatum (Buch.-Ham.) Sw.
Dendrobium nobile Lindl.
Dendrobium nobile var. nobile
Dendrobium ochreatum Lindl.
Dendrobium pauciflorum King & Pantl.
Dendrobium porphyrochilum Lindl.
Dendrobium stuposum Lindl.
Dendrobium transparens Wall. ex Lindl.
Dendrobium wardianum Warner

BOLIVIA/ BOLIVIE (LA) / BOLIVIA

Encyclia buchtienii Schltr.
Encyclia grammatoglossa (Rchb.f.) Dressler
Encyclia hartwegii (Lindl.) R.Vásquez & Dodson
Encyclia pflanzii Schltr.
Encyclia steinbachii Schltr.
Encyclia vespa (Vell.) Dressler
Encyclia yauperyensis (Barb.Rodr.) Porto & Brade

BRAZIL / BRÉSIL (LE) / BRASIL (EL)

Encyclia acuta Schltr.
Encyclia advena (Rchb.f.) Porto & Brade
Encyclia alagoenis (Pabst) Pabst
Encyclia albopurpurea (Barb.Rodr.) Porto & Brade
Encyclia alboxanthina Fowlie
Encyclia allemanii (Barb.Rodr.) Pabst
Encyclia allemanoides (Hoehne) Pabst

Encyclia amicta (Lindl. & Rchb.f.) Schltr.
Encyclia andrichii L.C.Menezes
Encyclia apuahuensis (Mansf.) Pabst
Encyclia argentinensis (Speg.) Hoehne
Encyclia baculus (Rchb.f.) Dressler & G.E.Pollard
Encyclia bracteata (Barb.Rodr.) Hoehne
Encyclia bragrançae Ruschi
Encyclia burle-marxii Pabst
Encyclia caetensis (Bicalho) Pabst
Encyclia calamaria (Lindl.) Pabst
Encyclia campos-portoi Pabst
Encyclia cardimii Pabst
Encyclia carpatiana (L.Lind.) Fowlie & Duv.
Encyclia chapadensis L.C.Menezes
Encyclia conchaechila (Barb.Rodr.) Porto & Brade
Encyclia dichroma (Lindl.) Schltr.
Encyclia doeringii Hoehne
Encyclia dutrai Pabst
Encyclia duveenii Pabst
Encyclia ensiformis (Ruiz & Pavon) Mansf.
Encyclia euosma Rchb.f.
Encyclia faresiana (Bicalho) Pabst
Encyclia fausta (Rchb.f. ex Cogn.) Pabst
Encyclia flabellifera Hoehne & Schltr.
Encyclia flava (Lindl.) Porto & Brade
Encyclia fowliei Duv.
Encyclia fragrans (Sw.) Lemée
Encyclia gallopavina (Rchb.f.) Porto & Brade
Encyclia ghillanyi Pabst
Encyclia glumacea (Lindl.) Pabst
Encyclia gonzalezii L.C.Menezes
Encyclia goyazensis L.C.Menezes
Encyclia granitica (Lindl.) Schltr.
Encyclia gravida (Lindl.) Schltr.
Encyclia hoehnei (Hawkes) Pabst
Encyclia hollandiae Fowlie
Encyclia huebneri Schltr.
Encyclia inversa (Lindl.) Pabst
Encyclia ionophlebium (Rchbf.) Dressler
Encyclia ionosma (Lindl.) Schltr.
Encyclia ivonae Carnevali & G.A. Romero
Encyclia kautskyi Pabst
Encyclia latipetala (C.Schweinf.) Pabst
Encyclia laxa Schltr.
Encyclia linearifolioides (Kraenzl.) Hoehne
Encyclia longifolia (Barb.Rodr.) Schltr.
Encyclia lutzenbergerii L.C.Menezes
Encyclia mapuerae Huber
Encyclia megalantha (Barb.Rodr.) Porto & Brade
Encyclia microxanthina Fowlie
Encyclia moojenii (Pabst) Pabst

Encyclia oncidioides (Lindl.) Schltr.
Encyclia osmantha (Barb.Rodr.) Schltr.
Encyclia oxiphylla Schltr.
Encyclia pachyantha (Lindl.) Hoehne
Encyclia papilio (Vell.) Pabst
Encyclia patens Hook.
Encyclia pauciflora (Barb.Rodr.) Porto & Brade
Encyclia pedra-azulensis L.C.Menezes
Encyclia peraltense (Ames) Withner
Encyclia pipio (Rchb.f.) Pabst
Encyclia punctifera (Rchb.f.) Pabst
Encyclia pygmaea (Hook.) Dressler
Encyclia randii (Barb.Rodr.) Porto & Brade
Encyclia regnelliana (Hoehne & Schltr.) Pabst
Encyclia rufa (Lindl.) Britton & Millsp.
Encyclia sessiflora (Edwall) Pabst
Encyclia schmidtii L.C.Menezes
Encyclia seidelii Pabst
Encyclia spiritusanctensis L.C.Menezes
Encyclia suzanensis (Hoehne) Pabst
Encyclia tarumana Schltr.
Encyclia tigrina (Linden ex Lindl.) Carnevali & Ramírez
Encyclia tripartita (Vell.) Hoehne
Encyclia unaensis Fowlie
Encyclia vellozoana Pabst
Encyclia vespa (Vell.) Dresller
Encyclia viridiflava L.C.Menzes
Encyclia viridiflora Hook.
Encyclia widgrenii (Lindl.) Pabst
Encyclia xerophytica Pabst
Encyclia xipheroides (Kraenzl.) Porto & Brade
Encyclia xuxiana Fowlie & Duv.
Encyclia yauperyensis (Barb.Rodr.) Porto & Brade

BRUNEI DARUSSALAM / BRUNÉI DARUSSALAM (LE) / BRUNEI DARUSSALAM

Dendrobium bostrychodes Rchb.f.
Dendrobium cinnabarinum Rchb.f.
Dendrobium cinnabarinum var. **cinnabarinum**
Dendrobium crumenatum Sw.
Dendrobium lobulatum Rolfe & J.J.Sm.
Dendrobium rosellum Ridl.
Dendrobium secundum (Blume) Lindl. ex Wall.

BURUNDI / BURUNDI (LE) / BURUNDI

Disa aconitoides Sond.
Disa aconitoides subsp. **aconitoides**
Disa cryptantha Summerh.
Disa eminii Kraenzl.

Disa erubescens Rendle
Disa erubescens subsp. **carsonii** (N.E.Br.) H.P.Linder
Disa erubescens subsp. **erubescens**
Disa hircicornis Rchb.f.
Disa ochrostachya Rchb.f.
Disa robusta N.E.Br.
Disa walleri Rchb.f.
Disa welwitschii Rchb.f.
Disa welwitschii subsp. **welwitschii**
Disa zombica N.E.Br.

CAMBODIA / CAMBODGE (LE) / CAMBOYA

Cymbidium aloifolium (L.) Sw.
Cymbidium bicolor subsp. **obtusum** Du Puy & P.J.Cribb
Cymbidium cyperifolium Wall. ex Lindl.
Cymbidium cyperifolium subsp. **arrogans** Du Puy & P.J.Cribb
Cymbidium dayanum Rchb.f.
Cymbidium ensifolium (L.) Sw.
Cymbidium finlaysonianum Lindl.
Dendrobium acinaciforme Roxb.
Dendrobium aloifolium (Blume) Rchb.f.
Dendrobium aphyllum (Roxb.) Fischer
Dendrobium crumenatum Sw.
Dendrobium crystallinum Rchb.f.
Dendrobium cumulatum Lindl.
Dendrobium delacourii Guill.
Dendrobium draconis Rchb.f.
Dendrobium incurvum Lindl.
Dendrobium intricatum Gagnep.
Dendrobium leonis (Lindl.) Rchb.f.
Dendrobium lomatochilum Seidenf.
Dendrobium microbolbon A.Rich.
Dendrobium nathanielis Rchb.f.
Dendrobium secundum (Blume) Lindl. ex Wall.
Dendrobium venustum Teijsm. & Binn.

CAMEROON / CAMEROUN (LE) / CAMERÚN (EL)

Disa equestris Rchb.f.
Disa erubescens Rendle
Disa erubescens subsp. **erubescens**
Disa hircicornis Rchb.f.
Disa nigerica Rolfe
Disa ochrostachya Rchb.f.
Disa perplexa H.P.Linder
Disa welwitschii Rchb.f.
Disa welwitschii subsp. **occultans** (Schltr.) H.P.Linder
Disa welwitschii subsp. **welwitschii**

CENTRAL AFRICAN REPUBLIC / RÉPUBLIQUE CENTRAFRICAINE (LA) / REPÚBLICA CENTROAFRICANA (LA)

Disa equestris Rchb.f.
Disa welwitschii Rchb.f.
Disa welwitschii subsp. **occultans** (Schltr.) H.P.Linder
Disa welwitschii subsp. **welwitschii**

CHINA / CHINE (LA) / CHINA

Cymbidium aloifolium (L.) Sw.
Cymbidium bicolor subsp. **obtusum** Du Puy & P.J.Cribb
Cymbidium cochleare Lindl.
Cymbidium cyperifolium Wall. ex Lindl.
Cymbidium dayanum Rchb.f.
Cymbidium defoliatum Y.S.Wu & S.C.Chen
Cymbidium eburneum Lindl.
Cymbidium elegans Lindl.
Cymbidium ensifolium (L.) Sw.
Cymbidium faberi Rolfe
Cymbidium faberi var. **faberi**
Cymbidium faberi var. **szechuanicum** (Y.S.Wu & S.C.Chen) Y.S.Wu & S.C.Chen
Cymbidium floribundum Lindl.
Cymbidium goeringii (Rchb.f.) Rchb.f.
Cymbidium goeringii var. **goeringii**
Cymbidium goeringii var. **longibracteatum** Y.S.Wu & S.C.Chen
Cymbidium goeringii var. **serratum** (Schltr.) Y.S.Wu & S.C.Chen
Cymbidium goeringii var. **tortisepalum** (Fukuy.) Y.S.Wu & S.C.Chen
Cymbidium gongshanense H.Li & G.H.Feng
Cymbidium hookerianum Rchb.f.
Cymbidium insigne Rolfe
Cymbidium iridioides D.Don
Cymbidium kanran Makino
Cymbidium lancifolium Hook.
Cymbidium lowianum (Rchb.f.) Rchb.f.
Cymbidium lowianum var. **iansonii** (Rolfe) P.J.Cribb & Du Puy
Cymbidium lowianum var. **lowianum**
Cymbidium macrorhizon Lindl.
Cymbidium maguanense F.Y.Liu
Cymbidium mastersii Griff. ex Lindl.
Cymbidium nanulum Y.S.Wu & S.C.Chen
Cymbidium qiubeiense K.M.Feng & H.Li
Cymbidium sinense (G.Jacks. & H.C.Andr.) Wild.
Cymbidium suavissimum Hort. ex C.H.Curtis
Cymbidium tigrinum Par. ex Hook
Cymbidium tracyanum L.Castle
Cymbidium wenshanense Y.S.Wu & F.Y.Liu
Cymbidium wilsonii (Rolfe) Rolfe
Dendrobium acinaciforme Roxb.
Dendrobium aduncum Wall. ex Lindl.
Dendrobium aphyllum (Roxb.) Fischer

Dendrobium bellatulum Rolfe
Dendrobium brymerianum Rchb.f.
Dendrobium capillipes Rchb.f.
Dendrobium cariniferum Rchb.f.
Dendrobium chagjiangense S.J. Cheng & C.Z. Tang
Dendrobium chrysanthum Wall.
Dendrobium chryseum Rolfe
Dendrobium chrysotoxum Lindl.
Dendrobium compactum Rolfe ex W.Hackett
Dendrobium crepidatum Lindl. & Paxton
Dendrobium crumenatum Sw.
Dendrobium crystallinum Rchb.f.
Dendrobium densiflorum Wall. ex Lindl.
Dendrobium dixanthum Rchb.f.
Dendrobium devonianum Paxton
Dendrobium dixanthum Rchb.f.
Dendrobium elipsophyllum Tang & Wang
Dendrobium exile Schltr.
Dendrobium falconeri Hook.
Dendrobium fimbriatum Hook.
Dendrobium findlayanum Pax. & Rchb.f.
Dendrobium flexicaule Z.H.Tsi, S.C.Sun & L.G.Xu
Dendrobium gibsonii Lindl.
Dendrobium gratiosissimum Rchb.f.
Dendrobium guangxiense S.J.Cheng & C.Z.Tang
Dendrobium hainanense Rolfe
Dendrobium hancockii Rolfe
Dendrobium harveyanum Rchb.f.
Dendrobium henryi Schltr.
Dendrobium hercoglossum Rchb.f.
Dendrobium heterocarpum Lindl.
Dendrobium hokerianum Lindl.
Dendrobium huoshannense C.Z.Tang & S.J.Cheng
Dendrobium infundibulum Lindl.
Dendrobium jenkinsii Wall. ex Lindl.
Dendrobium linawianum Rchb.f.
Dendrobium lituiflorum Lindl.
Dendrobium loddigessi Rolfe
Dendrobium lindleyi Steud.
Dendrobium lituiflorum Lindl.
Dendrobium loddigesii Rolfe
Dendrobium lohoense Tang & Wang
Dendrobium longicornu Wall. ex Lindl.
Dendrobium minutiflorum S.C.Chen & Z.H.Tsi
Dendrobium moniliforme (L.) Sw.
Dendrobium monticola Hunt & Summerh.
Dendrobium moschatum (Buch.-Ham.) Sw.
Dendrobium nobile Lindl.
Dendrobium nobile var. **nobile**
Dendrobium officinale Kimura & Migo
Dendrobium parciflorum Rchb.f. ex Lindl.

Dendrobium parishii Rchb.f.
Dendrobium pendulum Roxb.
Dendrobium porphyrochilum Lindl.
Dendrobium primulinum Lindl.
Dendrobium pseudotenellum Guill.
Dendrobium pulchellum Roxb. ex Lindl.
Dendrobium sinense Tang & Wang
Dendrobium strongylanthum Rchb.f.
Dendrobium stuposum Lindl.
Dendrobium sulcatum Lindl.
Dendrobium terminale Parish & Rchb.f.
Dendrobium thyrsiflorum Rchb.f. ex Andre
Dendrobium tosaense Makino
Dendrobium trigonopus Rchb.f.
Dendrobium wardianum Warner
Dendrobium williamsonii Day & Rchb.f.
Dendrobium wilsonii Rolfe
Dendrobium xichouense C.J.Cheng & C.Z.Tang
Dendrobium zhaojuense S.C.Sun & L.G.Xu

CHINA (TAIWAN)

Cymbidium cochleare Lindl.
Cymbidium dayanum Rchb.f.
Cymbidium ensifolium (L.) Sw.
Cymbidium faberi Rolfe
Cymbidium faberi var. faberi
Cymbidium floribundum Lindl.
Cymbidium goeringii (Rchb.f.) Rchb.f.
Cymbidium goeringii var. goeringii
Cymbidium goeringii var. serratum (Schltr.) Y.S.Wu & S.C.Chen
Cymbidium goeringii var. tortisepalum (Fukuy.) Y.S.Wu & S.C.Chen
Cymbidium kanran Makino
Cymbidium lancifolium Hook.
Cymbidium macrorhizon Lindl.
Cymbidium sinense (G.Jacks. & H.C.Andr.) Wild.
Dendrobium chryseum Rolfe
Dendrobium crumenatum Sw.
Dendrobium falconeri Hook.
Dendrobium furcatopedicellatum Hayata
Dendrobium leptocladum Hayata
Dendrobium linawianum Rchb.f.
Dendrobium miyakei Schltr.
Dendrobium moniliforme (L.) Sw.
Dendrobium nobile Lindl.
Dendrobium nobile var. nobile
Dendrobium somai Hayata
Dendrobium tosaense Makino
Dendrobium ventricosum Kraenzl.

COLOMBIA / COLOMBIE (LA) / COLOMBIA

Dracula alcithoe Luer & R.Escobar
Dracula amaliae Luer & R.Escobar
Dracula andreettae (Luer) Luer
Dracula anicula Luer & R.Escobar
Dracula anthracina Luer & R.Escobar
Dracula aphrodes Luer & R.Escobar
Dracula bella (Rchb.f.) Luer
Dracula bellerophon Luer & R.Escobar
Dracula benedictii (Rchb.f.) Luer
Dracula berthae Luer & R.Escobar
Dracula brangeri Luer
Dracula carcinopsis Luer & R.Escobar
Dracula chestertonii (Rchb.f.) Luer
Dracula chimaera (Rchb.f.) Luer
Dracula chiroptera Luer & Malo
Dracula × **circe** Luer & R.Escobar
Dracula citrina Luer & R.Escobar
Dracula cochliops Luer & R.Escobar
Dracula cutis-bufonis Luer & R.Escobar
Dracula decussata Luer & R.Escobar
Dracula diabola Luer & R.Escobar
Dracula diana Luer & R.Escobar
Dracula exasperata Luer & R.Escobar
Dracula felix (Luer) Luer
Dracula gigas (Luer & Andreetta) Luer
Dracula gorgona (Veitch) Luer & R.Escobar
Dracula gorgonella Luer & R.Escobar
Dracula hirtzii Luer
Dracula houtteana (Rchb.f.) Luer
Dracula inaequalis (Rchb.f.) Luer & R.Escobar
Dracula incognita Luer & R.Escobar
Dracula insolita Luer & R.Escobar
Dracula lehmanniana Luer & R.Escobar
Dracula lemurella Luer & R.Escobar
Dracula levii Luer
Dracula ligiae Luer & R.Escobar
Dracula mantissa Luer & R.Escobar
Dracula minax Luer & R.Escobar
Dracula niesseniae Ortiz.
Dracula nosferatu Luer & R.Escobar
Dracula nycterina (Rchb.f.) Luer
Dracula octavioi Luer & R.Escobar
Dracula ophioceps Luer & R.Escobar
Dracula orientalis Luer & R.Escobar
Dracula ortiziana Luer & R.Escobar
Dracula pholeodytes Luer & R.Escobar
Dracula × **pileus** Luer & R.Escobar
Dracula platycrater (Rchb.f.) Luer
Dracula posadarum Luer & R.Escobar
Dracula presbys Luer & R.Escobar

Dracula psittacina (Rchb.f.) Luer & R.Escobar
Dracula radiosa (Rchb.f.) Luer
Dracula × radio-syndactyla Luer
Dracula robledorum (P.Ortiz) Luer & R.Escobar
Dracula roezlii (Rchb.f.) Luer
Dracula sergioi Luer & R.Escobar
Dracula severa (Rchb.f.) Luer
Dracula sibundoyensis Luer & R.Escobar
Dracula syndactyla Luer
Dracula trichroma (Schltr.) Hermans
Dracula velutina (Rchb.f.) Luer
Dracula venefica Luer & R.Escobar
Dracula venosa (Rolfe) Luer
Dracula verticulosa Luer & R.Escobar
Dracula vespertilio (Rchb.f.) Luer
Dracula vinacea Luer & R.Escobar
Dracula vlad-tepes Luer & R.Escobar
Dracula wallisii (Rchb.f.) Luer
Dracula xenos Luer & R.Escobar
Encyclia abbreviata (Schltr.) Dressler
Encyclia alata (Bateman) Schltr.
Encyclia arminii (Rchb.f.) Carnevali & Ramírez
Encyclia asperirachis Garay
Encyclia atrorubens (Rolfe) Schltr.
Encyclia baculus (Rchb.f.) Dressler & G.E.Pollard
Encyclia ceratistes (Lindl.) Schltr.
Encyclia chacaoensis (Rchb.f.) Dressler & G.E.Pollard
Encyclia chimborazoensis (Schltr.) Dressler
Encyclia cochleata (L.) Lemée
Encyclia cordigera (Kunth) Dressler
Encyclia diota (Lindl.) Schltr.
Encyclia fragrans (Sw.) Lemée
Encyclia grammatoglossa (Rchb.f.) Dressler
Encyclia hartwegii (Rchb.f.) Pabst, Moutinho & A.V.Pinto
Encyclia ionophlebium (Rchb.f.) Dressler
Encyclia lambda (Linden & Rchb.f.) Dressler
Encyclia leucantha Schltr.
Encyclia lindenii (Lindl.) Carnevali & Ramírez
Encyclia livida (Lindl.) Dressler
Encyclia luteorosea (A.Rich. & H.Galeotti) Dressler & Pollard
Encyclia maderoi Schltr.
Encyclia pulcherrima Dodson & Bennett
Encyclia sceptra (Lindl.) Carnevali & Ramírez
Encyclia tigrina (Linden ex Lindl.) Carnevali & Ramírez
Encyclia vespa (Vell.) Dressler

CONGO (THE DEMOCRATIC REPUBLIC OF THE) / RÉPUBLIQUE DÉMOCRATIQUE DU CONGO (LA) / REPÚBLICA DEMOCRÁTICA DEL CONGO (LA)

Disa aconitoides Sond.

Disa aconitoides subsp. **aconitoides**
Disa aequiloba Summerh.
Disa caffra Bolus
Disa cryptantha Summerh.
Disa danielae Greerinck
Disa dichroa Summerh.
Disa englerana Kraenzl.
Disa equestris Rchb.f.
Disa erubescens Rendle
Disa erubescens subsp. **carsonii** (N.E.Br.) H.P.Linder
Disa erubescens subsp. **erubescens**
Disa fragrans Schltr.
Disa fragrans subsp. **deckenii** (Rchb.f.) H.P.Linder
Disa fragrans subsp. **fragrans**
Disa hircicornis Rchb.f.
Disa katangensis De Wild.
Disa miniata Summerh.
Disa nigerica Rolfe
Disa ochrostachya Rchb.f.
Disa ornithantha Schltr.
Disa perplexa H.P.Linder
Disa robusta N.E.Br.
Disa roeperocharoides Kraenzl.
Disa stairsii Kraenzl.
Disa verdickii De Wild.
Disa walleri Rchb.f.
Disa welwitschii Rchb.f.
Disa welwitschii subsp. **occultans** (Schltr.) H.P.Linder
Disa welwitschii subsp. **welwitschii**
Disa zombica N.E.Br.

COSTA RICA / COSTA RICA (LE) / COSTA RICA

Dracula carlueri Hermans & P.J.Cribb
Dracula erythrochaete (Rchb.f.) Luer
Dracula erythrochaete subsp. **astuta** (Rchb.f.) Luer
Dracula erythrochaete subsp. **santa-elenae** Hermans
Dracula pusilla (Rolfe) Luer
Dracula ripleyana Luer
Dracula vespertilio (Rchb.f.) Luer
Encyclia abbreviata (Schltr.) Dressler
Encyclia alata (Bateman) Schltr.
Encyclia baculus (Rchb.f.) Dressler & G.E.Pollard
Encyclia brassavolae (Rchb.f.) Dressler
Encyclia candollei (Lindl.) Schltr.
Encyclia ceratistes (Lindl.) Schltr.
Encyclia cordigera (Kunth) Dressler
Encyclia mooreana (Rolfe) Schltr.
Encyclia neurosa (Ames) Dressler & Pollard
Encyclia ochracea (Lindl.) Dressler
Encyclia prismatocarpa (Rchb.f.) Dressler

Encyclia spondiadum (Rchb.f.) Dressler
Encyclia stellata (Lindl.) Schltr.
Encyclia vagans (Ames) Dressler & G.E.Pollard
Encyclia varicosa Bateman ex Lindl.
Encyclia varicosa subsp. **varicosa**
Encyclia vespa (Vell.) Dressler

CÔTE D'IVOIRE / CÔTE D'IVOIRE (LA) / CÔTE D'IVOIRE (LA)

Disa welwitschii Rchb.f.
Disa welwitschii subsp. **occultans** (Schltr.) H.P.Linder
Disa welwitschii subsp. **welwitschii**

CUBA / CUBA / CUBA

Encyclia boothiana (Lindl.) Dressler
Encyclia boothiana subsp. **boothiana**
Encyclia bractescens (Lindl.) Hoehne
Encyclia cochleata (L.) Lemée
Encyclia nematocaulon (A.Rich.) Acuña
Encyclia polybulbon (Sw.) Dressler
Encyclia triangulifera (Rchb.f.) Acuña.

ECUADOR / EQUATEUR (L') / ECUADOR (EL)

Dracula alcithoe Luer & R.Escobar
Dracula andreettae (Luer) Luer
Dracula chiroptera Luer & Malo
Dracula cordobae Luer
Dracula dalessandroi Luer
Dracula dalstroemii Luer
Dracula deltoidea (Luer) Luer
Dracula dodsonii (Luer) Luer
Dracula fafnir Luer
Dracula felix (Luer) Luer
Dracula fuligifera Luer
Dracula gastrophora Luer & Hirtz
Dracula gigas (Luer & Andreetta) Luer
Dracula hawleyi Luer
Dracula hirsuta Luer & Andreetta
Dracula hirtzii Luer
Dracula lafleurii Luer & Dalström
Dracula levii Luer
Dracula lindstroemii Luer & Dalström
Dracula lotax (Luer) Luer
Dracula mantissa Luer & R.Escobar
Dracula marsupialis Luer & Hirtz
Dracula mopsus (F.C.Lehman & Kraenzl.) Luer
Dracula morleyi Luer & Dalström
Dracula navarroorum Luer & Hirtz

Dracula papillosa Luer & Dodson
Dracula polyphemus (Luer) Luer
Dracula portillae Luer & Andreetta
Dracula psyche (Luer) Luer
Dracula pubescens Luer & Dalström.
Dracula radiella Luer
Dracula radiosa (Rchb.f.) Luer
Dracula rezekiana Luer & Hawley
Dracula sibundoyensis Luer & R.Escobar
Dracula simia (Luer) Luer
Dracula sodiroi (Schltr.) Luer
Dracula trichroma (Schltr.) Hermans
Dracula trinympharum Luer
Dracula tubeana (Rchb.f.) Luer
Dracula ubangina Luer & Andreetta
Dracula vampira (Luer) Luer
Dracula venosa (Rolfe) Luer
Dracula vespertilio (Rchb.f.) Luer
Dracula wallisii (Rchb.f.) Luer
Dracula woolwardiae (Lehm.) Luer
Encyclia abbreviata (Schltr.) Dressler
Encyclia addenocarpon (La Llave & Lex.) Schltr.
Encyclia angustiloba Schltr.
Encyclia aspera (Lindl.) Schltr.
Encyclia chimborazoensis (Schltr.) Dressler
Encyclia chloroleuca (Hook.) Neum.
Encyclia grammatoglossa (Rchb.f.) Dressler
Encyclia hartwegii (Lindl.) R.Vásquez & Dodson
Encyclia livida (Lindl.) Dressler
Encyclia microtos (Rchb.f.) Hoehne
Encyclia naranjapatensis Dodson
Encyclia tigrina (Linden ex Lindl.) Carnevali & Ramírez
Encyclia vespa (Vell.) Dressler

EL SALVADOR / EL SALVADOR (L') / EL SALVADOR

Encyclia adenocarpon (La Llave & Lex.) Schltr.
Encyclia ambigua (Lindl.) Schltr.
Encyclia atrorubens (Rolfe) Schltr.
Encyclia baculus (Rchb.f.) Dressler & G.E.Pollard
Encyclia brassavolae (Rchb.f.) Dressler
Encyclia chondylobulbon (A.Rich. & H.Galeotti) Dressler & Pollard
Encyclia cochleata (L.) Lemée
Encyclia cordigera (Kunth) Dressler
Encyclia dickinsoniana (Withner) Withner
Encyclia diota (Lindl.) Schltr.
Encyclia glauca (Knowles & Westc.) Dressler & G.E.Pollard
Encyclia incumbens (Lindl.) Mabb.
Encyclia livida (Lindl.) Dressler
Encyclia luteorosea (A.Rich. & H.Galeotti) Dressler & Pollard
Encyclia michuacana (La Llave & Lex.) Schltr.

Encyclia mooreana (Rolfe) Schltr.
Encyclia nematocaulon (A.Rich.) Acuña
Encyclia ochracea (Lindl.) Dressler
Encyclia polybulbon (Sw.) Dressler
Encyclia pseudopygmaea (Finet) Dressler & G.E.Pollard
Encyclia radiata (Lindl.) Dressler
Encyclia rhynchophora (A.Rich. & H.G.Galeotti) Dressler
Encyclia selligera (Bateman ex Lindl.) Schltr.
Encyclia trachycarpa (Lindl. ex Benth) Schltr.
Encyclia vagans (Ames) Dressler & G.E.Pollard
Encyclia varicosa Bateman ex Lindl.
Encyclia varicosa subsp. varicosa
Encyclia vitellina (Lindl.) Dressler

ETHIOPIA / ETHIOPIE (L') / ETIOPA

Disa aconitoides Sond.
Disa aconitoides subsp. aconitoides
Disa aconitoides subsp. goetzeana (Kraenzl.) H.P.Linder
Disa fragrans Schltr.
Disa fragrans subsp. deckenii (Rchb.f.) H.P.Linder
Disa fragrans subsp. fragrans
Disa pulchella Hochst. ex A.Rich.
Disa scutellifera A.Rich.

FIJI / FIDJI (LES) / FIJI

Dendrobium catillare Rchb.f.
Dendrobium delicatulum Kraenzl.
Dendrobium delicatulum subsp. delicatulum
Dendrobium hornei S.Moore ex Baker
Dendrobium macrophyllum A.Rich.
Dendrobium macrophyllum var. macrophyllum
Dendrobium macropus (Endl.) Rchb.f. ex Lindl.
Dendrobium masarangense Schltr
Dendrobium masarangense subsp. masarangense
Dendrobium mohlianum Rchb.f.
Dendrobium prasinum Lindl.
Dendrobium spathulatum L.O.Williams
Dendrobium taveuniense Dauncey & P.J.Cribb
Dendrobium tokai Rchb.f.
Dendrobium vagans Schltr.

GUATEMALA / GUATEMALA (LE) / GUATEMALA

Dracula pusilla (Rolfe) Luer
Encyclia abbreviata (Schltr.) Dressler
Encyclia adenocarpon (La Llave & Lex.) Schltr.
Encyclia aenicta Dressler & G.E.Pollard
Encyclia alata (Bateman) Schltr.
Encyclia ambigua (Lindl.) Schltr.

Encyclia asperula Dressler & G.E.Pollard
Encyclia baculus (Rchb.f.) Dressler & G.E.Pollard
Encyclia bractescens (Lindl.) Hoehne
Encyclia candollei (Lindl.) Schltr.
Encyclia cepiforme Hooker
Encyclia chondylobulbon (A.Rich. & H.Galeotti) Dressler & Pollard
Encyclia dickinsoniana (Withner) Withner
Encyclia glauca (Knowles & Westc.) Dressler & G.E.Pollard
Encyclia incumbens (Lindl.) Mabb.
Encyclia maculosa (Ames, F.T.Hubb. & C.Schweinf.) Hoehne
Encyclia michuacana (La Llave & Lex.) Schltr.
Encyclia nematocaulon (A.Rich.) Acuña
Encyclia neurosa (Ames) Dressler & Pollard
Encyclia panthera (Rchb.f.) Schltr.
Encyclia papillosa (Bateman) Aguirre-Olav.
Encyclia polybulbon (Sw.) Dressler
Encyclia pseudopygmaea (Finet) Dressler & G.E.Pollard
Encyclia radiata (Lindl.) Dressler
Encyclia rhynchophora (A.Rich. & H.G.Galeotti) Dressler
Encyclia selligera (Lindl.) Schltr.
Encyclia tuerckheimii Schltr.
Encyclia vagans (Ames) Dressler & G.E.Pollard
Encyclia virens (Lindl.) Schltr.
Encyclia vitellina (Lindl.) Dressler

GUINEA / GUINÉE (LA) / GUINEA

Disa welwitschii Rchb.f.
Disa welwitschii subsp. occultans (Schltr.) H.P.Linder
Disa welwitschii subsp. welwitschii

GUYANA / GUYANA (LE) / GUYANA

Encyclia chloroleuca (Hook.) Neum.
Encyclia cochleata (L.) Lemée
Encyclia fragrans (Sw.) Lemée
Encyclia granitica (Lindl.) Schltr.
Encyclia latipetala (C.Schweinf.) Pabst
Encyclia luteorosea (A.Rich. & H.Galeotti) Dressler & Pollard
Encyclia ochracea (Lindl.) Dressler
Encyclia tigrina (Linden ex Lindl.) Carnevali & Ramírez
Encyclia trachycarpa (Lindl.) Schltr.
Encyclia varicosa Bateman ex Lindl.
Encyclia varicosa subsp. varicosa
Encyclia vespa (Vell.) Dressler

HONDURAS / HONDURAS (LE) / HONDURAS

Encyclia abbreviata (Schltr.) Dressler
Encyclia adenocarpon (La Llave & Lex.) Schltr.
Encyclia aenicta Dressler & G.E.Pollard

Encyclia alata (Bateman) Schltr.
Encyclia ambigua (Lindl.) Schltr.
Encyclia baculus (Rchb.f.) Dressler & G.E.Pollard
Encyclia boothiana (Lindl.) Dressler
Encyclia boothiana subsp. boothiana
Encyclia bractescens (Lindl.) Hoehne
Encyclia ceratistes (Lindl.) Schltr.
Encyclia dickinsoniana (Withner) Withner
Encyclia glauca (Knowles & Westc.) Dressler & G.E.Pollard
Encyclia guatemalensis (Klotzsch) Withner
Encyclia luteorosea (A.Rich. & H.Galeotti) Dressler & Pollard
Encyclia michuacana (La Llave & Lex.) Schltr.
Encyclia nematocaulon (A.Rich.) Acuña
Encyclia neurosa (Ames) Dressler & Pollard
Encyclia ochracea (Lindl.) Dressler
Encyclia papillosa (Bateman) Aguirre-Olav.
Encyclia polybulbon (Sw.) Dressler
Encyclia radiata (Lindl.) Dressler
Encyclia rhynchophora (A.Rich. & H.G.Galeotti) Dressler
Encyclia selligera (Lindl.) Schltr.
Encyclia vagans (Ames) Dressler & G.E.Pollard

INDIA / INDE (L') / INDIA (LA)

Cymbidium aloifolium (L.) Sw.
Cymbidium bicolor Lindl.
Cymbidium bicolor subsp. bicolor
Cymbidium bicolor subsp. obtusum Du Puy & P.J.Cribb
Cymbidium cochleare Lindl.
Cymbidium cyperifolium Wall. ex Lindl.
Cymbidium cyperifolium subsp. cyperifolium
Cymbidium dayanum Rchb.f.
Cymbidium devonianum Paxton
Cymbidium eburneum Lindl.
Cymbidium elegans Lindl.
Cymbidium ensifolium (L.) Sw.
Cymbidium ensifolium subsp. haematodes (Lindl.) Du Puy & P.J.Cribb
Cymbidium faberi Rolfe
Cymbidium faberi var. szechuanicum (Y.S.Wu & S.C.Chen) Y.S.Wu & S.C.Chen
Cymbidium goeringii (Rchb.f.) Rchb.f.
Cymbidium goeringii var. goeringii
Cymbidium hookerianum Rchb.f.
Cymbidium iridioides D.Don
Cymbidium lancifolium Hook.
Cymbidium macrorhizon Lindl.
Cymbidium mastersii Griff. ex Lindl.
Cymbidium munronianum King & Pantl.
Cymbidium sinense (G.Jacks. & H.C.Andr.) Wild.
Cymbidium tigrinum Parish ex Hook.
Cymbidium whiteae King & Pantl.
Dendrobium acinaciforme Roxb.

Dendrobium aduncum Wall. ex Lindl.
Dendrobium amoenum Lindl.
Dendrobium anceps Sw.
Dendrobium anosmum Lindl.
Dendrobium aphyllum (Roxb.) Fischer
Dendrobium aqueum Lindl.
Dendrobium barbatulum Lindl.
Dendrobium bellatulum Rolfe
Dendrobium bensoniae Rchb.f.
Dendrobium bicameratum Lindl.
Dendrobium brymerianum Rchb.f.
Dendrobium brymerianum var. **brymerianum**
Dendrobium capillipes Rchb.f.
Dendrobium cariniferum Rchb.f.
Dendrobium chrysanthum Wall.
Dendrobium chryseum Rolfe
Dendrobium chrysocrepis Parish & Rchb.f.
Dendrobium chrysotoxum Lindl.
Dendrobium crepidatum Lindl. & Paxton
Dendrobium cretaceum Lindl.
Dendrobium crumenatum Sw.
Dendrobium cumulatum Lindl.
Dendrobium curviflorum Rolfe
Dendrobium darjeelingensis Pradhan
Dendrobium densiflorum Wall. ex Lindl.
Dendrobium denudans D.Don
Dendrobium devonianum Paxton
Dendrobium dickasonii L.O.Williams
Dendrobium draconis Rchb.f.
Dendrobium eriiflorum Griff.
Dendrobium falconeri Hook.
Dendrobium farmeri Paxton
Dendrobium fimbriatum Hook.
Dendrobium formosum Roxb. ex Lindl.
Dendrobium gibsonii Lindl.
Dendrobium graminifolium Lindl.
Dendrobium grande Hook.f.
Dendrobium gratiosissimum Rchb.f.
Dendrobium griffithianum Lindl.
Dendrobium herbaceum Lindl.
Dendrobium hercoglossom Rchb.f.
Dendrobium heterocarpum Lindl.
Dendrobium heyneanum Lindl.
Dendrobium hookerianum Lindl.
Dendrobium infundibulum Lindl.
Dendrobium jenkinsii Wall. ex Lindl.
Dendrobium lindleyi Steud.
Dendrobium lituiflorum Lindl.
Dendrobium longicornu Wall. ex Lindl.
Dendrobium macrostachyum Lindl.
Dendrobium mannii Ridl.

Dendrobium microbolbon A.Rich.
Dendrobium monticola Hunt & Summerh.
Dendrobium moschatum (Buch.-Ham.) Sw.
Dendrobium nathanielis Rchb.f.
Dendrobium nobile Lindl.
Dendrobium nobile var. nobile
Dendrobium ochreatum Lindl.
Dendrobium ovatum (Willd.) Kraenzl.
Dendrobium palpebrae Lindl.
Dendrobium parciflorum Rchb. ex Lindl.
Dendrobium parishii Rchb.f.
Dendrobium pauciflorum King & Pantl.
Dendrobium peguanum Lindl.
Dendrobium pendulum Roxb.
Dendrobium podagraria Hook.f.
Dendrobium porphyrochilum Lindl.
Dendrobium primulinum Lindl.
Dendrobium pulchellum Roxb. ex Lindl.
Dendrobium pychnostachyum Lindl.
Dendrobium secundum (Blume) Lindl. ex Wall.
Dendrobium stuposum Lindl.
Dendrobium sulcatum Lindl.
Dendrobium terminale Parish & Rchb.f.
Dendrobium thyrsiflorum Rchb.f. ex Andre
Dendrobium tortile Lindl.
Dendrobium transparens Wall. ex Lindl.
Dendrobium wardianum Warner
Dendrobium wattii (Hook.f.) Rchb.f.
Dendrobium williamsonii Day & Rchb.f.

INDONESIA / INDONÉSIA (L') / INDONESIA

Cymbidium aloifolium (L.) Sw.
Cymbidium atropurpureum (Lindl.) Rolfe
Cymbidium bicolor Lindl.
Cymbidium bicolor subsp. pubescens (Lindl.) Du Puy & P.J.Cribb
Cymbidium chloranthum Lindl.
Cymbidium dayanum Rchb.f.
Cymbidium ensifolium (L.) Sw.
Cymbidium ensifolium subsp. haematodes (Lindl.) Du Puy & P.J.Cribb
Cymbidium finlaysonianum Lindl.
Cymbidium hartinahianum J.B.Comber & Nasution
Cymbidium lancifolium Hook.
Cymbidium roseum J.J.Sm.
Cymbidium sigmoideum J.J.Sm.
Dendrobium aciculare Lindl.
Dendrobium acutimentum J.J.Sm.
Dendrobium acutisepalum J.J.Sm.
Dendrobium affine (Decne) Steud.
Dendrobium alderwereltianum J.J.Sm.
Dendrobium aloifolium (Blume) Rchb.f.

Dendrobium amblyogenium Schltr.
Dendrobium amphigenyum Ridl.
Dendrobium angiense J.J.Sm.
Dendrobium angustiflorum J.J.Sm.
Dendrobium annae J.J.Sm.
Dendrobium anosmum Lindl.
Dendrobium antennatum Lindl.
Dendrobium anthrene Ridl.
Dendrobium aphanochilum Kraenzl
Dendrobium arcuatum J.J.Sm.
Dendrobium aries J.J.Sm.
Dendrobium aristeferum J.J.Sm.
Dendrobium asphale Rchb.f.
Dendrobium atavus J.J.Sm.
Dendrobium atjehense J.J.Sm.
Dendrobium aurantiroseum P.Royen ex T.M.Reeve
Dendrobium babiense J.J.Sm.
Dendrobium bicaudatum Reinw. ex Lindl.
Dendrobium bicornutum Schltr.
Dendrobium bifalce Lindl.
Dendrobium bigibbum Lindl.
Dendrobium biloculare J.J.Sm.
Dendrobium blumei Lindl.
Dendrobium boumaniae J.J.Sm.
Dendrobium brachycentrum Ridl.
Dendrobium bracteosum Rchb.f.
Dendrobium brevicaule Rolfe
Dendrobium brevicaule subsp. **brevicaule**
Dendrobium brevicaule subsp. **calcarium** (J.J.Sm.) T.M.Reeve & P.Woods
Dendrobium calceolum Roxb.
Dendrobium caliculimentum R.S.Rogers
Dendrobium calophyllum Rchb.f.
Dendrobium calyptratum J.J.Sm.
Dendrobium capituliflorum Rolfe
Dendrobium capra J.J.Sm.
Dendrobium cinereum J.J.Sm.
Dendrobium cochliodes Schltr.
Dendrobium compressimentum J.J.Sm.
Dendrobium conanthum Schltr.
Dendrobium conicum J.J.Sm.
Dendrobium constrictum J.J.Sm.
Dendrobium convexipes J.J.Sm.
Dendrobium corallorhizon J.J.Sm.
Dendrobium crenatifolium J.J.Sm.
Dendrobium croceocentrum J.J.Sm.
Dendrobium crucilabre J.J.Sm.
Dendrobium crumenatum Sw.
Dendrobium cruttwellii T.M.Reeve
Dendrobium cuculliferum J.J.Sm.
Dendrobium cumulatum Lindl.
Dendrobium curvimentum J.J.Sm.

Dendrobium cuthbertsonii F.Muell.
Dendrobium cyanocentrum Schltr.
Dendrobium cylindricum J.J.Sm.
Dendrobium cymbulipes J.J.Sm.
Dendrobium dekockii J.J.Sm.
Dendrobium delicatulum Kraenzl.
Dendrobium delicatulum subsp. **delicatulum**
Dendrobium delicatulum subsp. **parvulum** (Rolfe) T.M.Reeve & P.Woods
Dendrobium dendrocolloides J.J.Sm.
Dendrobium devosianum J.J.Sm.
Dendrobium dillonianum A.D.Hawkes & A.H.Heller
Dendrobium endertii J.J.Sm.
Dendrobium eriiflorum Griff.
Dendrobium erosum (Blume) Lindl.
Dendrobium eximium Schltr.
Dendrobium fimbrilabium J.J.Sm.
Dendrobium finisterrae Schltr.
Dendrobium fruticicola J.J.Sm.
Dendrobium fulgidum Schltr.
Dendrobium fulgidum var. **fulgidum**
Dendrobium fulgidum var. **maritimum** (J.J.Sm.) Dauncey
Dendrobium fulminicaule J.J.Sm.
Dendrobium gemellum Lindl.
Dendrobium glaucoviride J.J.Sm.
Dendrobium glomeratum Rolfe
Dendrobium gracile (Blume) Lindl.
Dendrobium grande Hook.f.
Dendrobium grastidioides J.J.Sm.
Dendrobium grootingsii J.J.Sm.
Dendrobium habbemense Van Royen
Dendrobium hallieri J.J.Sm.
Dendrobium hamiferum P.J.Cribb
Dendrobium hasseltii (Blume) Lindl.
Dendrobium hellwigianum Kraenzl.
Dendrobium hendersonii A.D.Hawkes & A.H.Heller
Dendrobium heterocarpum Lindl.
Dendrobium hymenophyllum Lindl.
Dendrobium hymenopterum Hook.f.
Dendrobium igneoniveum J.J.Sm.
Dendrobium incurvociliatum J.J.Sm.
Dendrobium indivisum (Blume) Miq.
Dendrobium indivisum var. **indivisum**
Dendrobium inflatum Rolfe
Dendrobium informe J.J.Sm.
Dendrobium infractum J.J.Sm.
Dendrobium jabiense J.J.Sm.
Dendrobium jacobsonii J.J.Sm.
Dendrobium johannis Rchb.f.
Dendrobium johnsoniae F.Muell.
Dendrobium kauldorumii T.M.Reeve
Dendrobium keytsianum J.J.Sm.

Dendrobium kiauense Ames & C.Schweinf.
Dendrobium klabatense Schltr.
Dendrobium korthalsii J.J.Sm.
Dendrobium kruiense J.J.Sm.
Dendrobium lamellatum (Blume) Lindl.
Dendrobium lamelluliferum J.J.Sm.
Dendrobium lamii J.J.Sm.
Dendrobium lampongense J.J.Sm.
Dendrobium lancifolium A.Rich.
Dendrobium lancilabium J.J.Sm.
Dendrobium lancilobum J.J.Wood
Dendrobium lanyaiae Seidenf.
Dendrobium lasianthera J.J.Sm.
Dendrobium laurensii J.J.Sm.
Dendrobium laxiflorum J.J.Sm.
Dendrobium leonis (Lindl.) Rchb.f.
Dendrobium leporinum J.J.Sm.
Dendrobium leucohybos Schltr.
Dendrobium linearifolium Teijsm. & Binn.
Dendrobium linguella Rchb.f.
Dendrobium lobatum (Blume) Miq.
Dendrobium lobulatum Rolfe & J.J.Sm.
Dendrobium lowii Lindl.
Dendrobium macrifolium J.J.Sm.
Dendrobium macrophyllum A.Rich.
Dendrobium macrophyllum var. macrophyllum
Dendrobium macrophyllum var. subvelutinum J.J.Sm.
Dendrobium maierae J.J.Sm.
Dendrobium malvicolor Ridl.
Dendrobium masarangense Schltr
Dendrobium masarangense subsp. chlorinum Ridl.
Dendrobium masarangense subsp. masarangense
Dendrobium masarangense var. theionanthum (Schltr.) T.M.Reeve & P.Woods
Dendrobium militare P.J.Cribb
Dendrobium mirbelianum Gaudich.
Dendrobium molle J.J.Sm.
Dendrobium montanum J.J.Sm.
Dendrobium mutabile (Blume) Lindl.
Dendrobium nebularum Schltr.
Dendrobium nieuwenhuisii J.J.Sm.
Dendrobium nudum (Blume) Lindl.
Dendrobium obcordatum J.J.Sm.
Dendrobium obtusum Schltr.
Dendrobium odoardii Kraenzl.
Dendrobium oliganthum Schltr.
Dendrobium oreogenum Schltr.
Dendrobium ovipostoriferum J.J.Sm.
Dendrobium paathii J.J.Sm.
Dendrobium paniferum J.J.Sm.
Dendrobium papilioniferum J.J.Sm.
Dendrobium papilioniferum var. ephemerum J.J.Sm.

Dendrobium papilioniferum var. **papilioniferum**
Dendrobium papuanum J.J.Sm.
Dendrobium parvifolium J.J.Sm.
Dendrobium paspalifolium J.J.Sm.
Dendrobium pedicellatum J.J.Sm.
Dendrobium percnanthum Rchb.f.
Dendrobium planibulbe Lindl.
Dendrobium pogoniates Rchb.f.
Dendrobium praetermissum Dauncey
Dendrobium prianganense J.J.Wood
Dendrobium pseudoconanthum J.J.Sm.
Dendrobium pseudoglomeratum T.M.Reeve & J.J.Woods
Dendrobium pseudopeloricum J.J.Sm.
Dendrobium puberilingue J.J.Sm.
Dendrobium purpureiflorum J.J.Sm.
Dendrobium purpureum Roxb.
Dendrobium purpureum subsp. **candidulum** (Rchb.f.) Dauncey & P.J.Cribb
Dendrobium purpureum subsp. **purpureum**
Dendrobium quadriquetrum J.J.Sm.
Dendrobium rachmatii J.J.Sm.
Dendrobium rantii J.J.Sm.
Dendrobium rappardii J.J.Sm.
Dendrobium rariflorum J.J.Sm.
Dendrobium rarum Schltr.
Dendrobium rarum var. **miscegeneum** Dauncey
Dendrobium rarum var. **rarum**
Dendrobium reflexibarbatulum J.J.Sm.
Dendrobium reflexitepalum J.J.Sm.
Dendrobium rhodostele Ridl.
Dendrobium rhomboglossum J.J.Sm.
Dendrobium rigidifolium Rolfe
Dendrobium rigidum R.Br.
Dendrobium rindjaniense J.J.Sm.
Dendrobium riparium J.J.Sm.
Dendrobium rosellum Ridl.
Dendrobium rupestre J.J.Sm.
Dendrobium rutriferum Rchb.f.
Dendrobium ruttenii J.J.Sm.
Dendrobium sagittatum J.J.Sm.
Dendrobium salmoneum Schltr.
Dendrobium sambasanum J.J.Sm.
Dendrobium sanguinolentum Lindl.
Dendrobium schulleri J.J.Sm.
Dendrobium scabrifolium Ridl.
Dendrobium secundum (Blume) Lindl. ex Wall.
Dendrobium seranicum J.J.Sm.
Dendrobium setifolium Ridl.
Dendrobium sidikalangense Dauncey
Dendrobium simplex J.J.Sm.
Dendrobium singkawangense J.J.Sm.
Dendrobium smillieae F.Muell.

Dendrobium smithianum Schltr.
Dendrobium spathilingue J.J.Sm.
Dendrobium spectabile (Blume) Miq.
Dendrobium spegidoglossum Rchb.f.
Dendrobium stellare Dauncey
Dendrobium stratiotes Rchb.f.
Dendrobium strebloceras Rchb.f.
Dendrobium strepsiceros J.J.Sm.
Dendrobium stuartii F.M.Bailey
Dendrobium subacaule Reinw. ex Lindl.
Dendrobium subclausum Rolfe
Dendrobium subclausum var. **subclausum**
Dendrobium subquadratum J.J.Sm.
Dendrobium subuliferum J.J.Sm.
Dendrobium sulphureum Schltr.
Dendrobium sulphureum var. **cellulosum** (J.J.Sm.) T.M.Reeve & P.Woods
Dendrobium sulphureum var. **sulphureum**
Dendrobium takahashii Carr
Dendrobium taurulinum J.J.Sm.
Dendrobium tenellum (Blume) Lindl.
Dendrobium tenue J.J.Sm.
Dendrobium terrestre J.J.Sm.
Dendrobium tetrachromum Rchb.f.
Dendrobium tetralobum Schltr.
Dendrobium tetrodon Rchb.f. ex Lindl.
Dendrobium thyrsodes Rchb.f.
Dendrobium torricellense Schltr.
Dendrobium transtilliferum J.J.Sm.
Dendrobium trichostomum Rchb.f. ex Oliver
Dendrobium tricuspe (Blume) Lindl.
Dendrobium trilamellatum J.J.Sm.
Dendrobium truncatum Lindl.
Dendrobium tubiflorum J.J.Sm.
Dendrobium uliginosum J.J.Sm.
Dendrobium uncatum Lindl.
Dendrobium uncipes J.J.Sm.
Dendrobium vannouhuysii J.J.Sm.
Dendrobium ventrilabium J.J.Sm.
Dendrobium vexillarius J.J.Sm.
Dendrobium vexillarius var. **microblepharum** (Schltr.) T.M.Reeve & P.Woods
Dendrobium vexillarius var. **retroflexum** (J.J.Sm.) T.M.Reeve & P.Woods
Dendrobium vexillarius var. **uncinatum** (Schltr.) T.M.Reeve & P.Woods
Dendrobium vexillarius var. **vexillarius**
Dendrobium violaceoflavens J.J.Sm.
Dendrobium violaceum Kraenzl.
Dendrobium violaceum subsp. **cyperifolium** (Schltr.) T.M.Reeve & P.Woods
Dendrobium violaceum subsp. **violaceum**
Dendrobium violascens J.J.Sm.
Dendrobium viriditepalum J.J.Sm.
Dendrobium wentianum J.J.Sm.
Dendrobium wisselense P.J.Cribb

Dendrobium woluense J.J.Sm.
Dendrobium womersleyi T.M.Reeve
Dendrobium womersleyi var. womersleyi
Dendrobium xanthoacron Schltr.

JAMAICA / JAMAÏQUE (LA) / JAMAICA

Encyclia fragrans (Sw.) Lemée
Encyclia polybulbon (Sw.) Dressler
Encyclia vespa (Vell.) Dressler

JAPAN / JAPON (LE) / JAPÓN (EL)

Cymbidium dayanum Rchb.f.
Cymbidium ensifolium (L.) Sw.
Cymbidium ensifolium subsp. ensifolium
Cymbidium goeringii (Rchb.f.) Rchb.f.
Cymbidium goeringii var. goeringii
Cymbidium goeringii var. serratum (Schltr.) Y.S.Wu & S.C.Chen
Cymbidium kanran Makino
Cymbidium lancifolium Hook.
Cymbidium macrorhizon Lindl.
Cymbidium sinense (G.Jacks. & H.C.Andr.) Wild.
Dendrobium minutiflorum S.C.Chen & Z.H.Tsi
Dendrobium moniliforme (L.) Sw.
Dendrobium okinawense Hatusima
Dendrobium tosaense Makino

KENYA / KENYA (LE) / KENYA

Disa aconitoides Sond.
Disa aconitoides subsp. aconitoides
Disa aconitoides subsp. goetzeana (Kraenzl.) H.P.Linder
Disa erubescens Rendle
Disa erubescens subsp. erubescens
Disa fragrans Schltr.
Disa fragrans subsp. deckenii (Rchb.f.) H.P.Linder
Disa fragrans subsp. fragrans
Disa hircicornis Rchb.f.
Disa ochrostachya Rchb.f.
Disa perplexa H.P.Linder
Disa scutellifera A.Rich.
Disa stairsii Kraenzl.
Disa welwitschii Rchb.f.
Disa welwitschii subsp. occultans (Schltr.) H.P.Linder
Disa welwitschii subsp. welwitschii

KOREA (THE REPUBLIC OF) / RÉPUBLIQUE DE CORÉE (LA) / REPÚBLICA DE COREA (LA)

Cymbidium goeringii (Rchb.f.) Rchb.f.
Cymbidium goeringii var. goeringii

Cymbidium kanran Makino
Dendrobium moniliforme (L.) Sw.

LAO PEOPLE'S DEMOCRATIC REPUBLIC (THE) / RÉPUBLIQUE DÉMOCRATIQUE POPULAIRE LAO (LA) / REPÚBLICA DEMOCRÁTICA POPULAR LAO (LA)

Cymbidium aloifolium (L.) Sw.
Cymbidium bicolor subsp. obtusum Du Puy & P.J.Cribb
Cymbidium dayanum Rchb.f.
Cymbidium ensifolium (L.) Sw.
Cymbidium lancifolium Hook.
Cymbidium macrorhizon Lindl.
Dendrobium acinaciforme Roxb.
Dendrobium aduncum Wall. ex Lindl.
Dendrobium aloifolium (Blume) Rchb.f.
Dendrobium anceps Sw.
Dendrobium anosmum Lindl.
Dendrobium aphyllum (Roxb.) Fischer
Dendrobium bellatulum Rolfe
Dendrobium brymerianum Rchb.f.
Dendrobium brymerianum var. brymerianum
Dendrobium capillipes Rchb.f.
Dendrobium cariniferum Rchb.f.
Dendrobium chrysanthum Wall.
Dendrobium chryseum Rolfe
Dendrobium chrysotoxum Lindl.
Dendrobium crepidatum Lindl. & Paxton
Dendrobium cretaceum Lindl.
Dendrobium crumenatum Sw.
Dendrobium crystallinum Rchb.f.
Dendrobium cumulatum Lindl.
Dendrobium delacourii Guill.
Dendrobium deltatum Seidnf.
Dendrobium densiflorum Wall. ex Lindl.
Dendrobium dixanthum Rchb.f.
Dendrobium draconis Rchb.f.
Dendrobium farmeri Paxton
Dendrobium fimbriatum Hook.
Dendrobium findlayanum Parish & Rchb.f.
Dendrobium friedericksianum Rchb.f.
Dendrobium gratiosissimum Rchb.f.
Dendrobium hercoglossum Rchb.f.
Dendrobium heterocarpum Lindl.
Dendrobium indivisum (Blume) Miq.
Dendrobium indivisum var. indivisum
Dendrobium infundibulum Lindl.
Dendrobium jenkinsii Wall. ex Lindl.
Dendrobium lamellatum (Blume) Lindl.
Dendrobium leonis (Lindl.) Rchb.f.
Dendrobium lindleyi Steud.

Dendrobium lituiflorum Lindl.
Dendrobium loddigesii Rolfe
Dendrobium mannii Ridl.
Dendrobium moschatum (Buch.-Ham.) Sw.
Dendrobium multilineatum Kerr
Dendrobium nathanielis Rchb.f.
Dendrobium nobile Lindl.
Dendrobium nobile var. nobile
Dendrobium ochreatum Lindl.
Dendrobium pachyglossum Par & Rchb.f.
Dendrobium palpebrae Lindl.
Dendrobium parcifolium Rchb.ex Lindl.
Dendrobium parishii Rchb.f.
Dendrobium pendulum Roxb.
Dendrobium porphyrophyllum Guill.
Dendrobium primulinum Lindl.
Dendrobium pulchellum Roxb. ex Lindl.
Dendrobium scabrilingue Lindl.
Dendrobium secundum (Blume) Lindl. ex Wall.
Dendrobium senile Parish & Rchb.f.
Dendrobium signatum Rchb.f.
Dendrobium sulcatum Lindl.
Dendrobium thyrsiflorum Rchb.f. ex Andre
Dendrobium tortile Lindl.
Dendrobium trigonopus Rchb.f.
Dendrobium unicum Seidenf.
Dendrobium venustum Teijsm. & Binn.
Dendrobium virgineum Rchb.f.
Dendrobium wattii (Hook.f.) Rchb.f.

LESOTHO / LESOTHO (LE) / LESOTHO

Disa basutorum Schltr.
Disa cephalotes Rchb.f.
Disa cephalotes subsp. cephalotes
Disa cephalotes subsp. frigida (Schltr.) H.P.Linder
Disa chrysostachya Sw.
Disa cooperi Rchb.f.
Disa cornuta (L.) Sw.
Disa crassicornis Lindl.
Disa fragrans Schltr.
Disa fragrans subsp. fragrans
Disa galpinii Rolfe
Disa nervosa Lindl.
Disa oreophila Bolus
Disa oreophila subsp. oreophila
Disa porrecta Sw.
Disa saxicola Schltr.
Disa stachyoides Rchb.f.
Disa stricta Sond.
Disa tysonii Bolus

Disa versicolor Rchb.f.

LIBERIA / LIBÉRIA (LE) / LIBERIA

Disa welwitschii Rchb.f.
Disa welwitschii subsp. **occultans** (Schltr.) H.P.Linder
Disa welwitschii subsp. **welwitschii**

MADAGASCAR / MADAGASCAR / MADAGASCAR

Disa andringitrana Schltr.
Disa buchenaviana Kraenzl.
Disa caffra Bolus
Disa incarnata Lindl.

MALAWI / MALAWI (LE) / MALAWI

Disa aconitoides Sond.
Disa aconitoides subsp. **aconitoides**
Disa aconitoides subsp. **concinna** (N.E.Br.) H.P.Linder
Disa aperta N.E.Br.
Disa caffra Bolus
Disa celata Summerh.
Disa englerana Kraenzl.
Disa equestris Rchb.f.
Disa erubescens Rendle
Disa erubescens subsp. **carsonii** (N.E.Br.) H.P.Linder
Disa erubescens subsp. **erubescens**
Disa fragrans Schltr.
Disa fragrans subsp. **fragrans**
Disa hircicornis Rchb.f.
Disa miniata Summerh.
Disa nyikensis H.P.Linder
Disa ochrostachya Rchb.f.
Disa ornithantha Schltr.
Disa perplexa H.P.Linder
Disa robusta N.E.Br.
Disa rungweensis Schltr.
Disa satyriopsis Kraenzl.
Disa saxicola Schltr.
Disa ukingensis Schltr.
Disa walleri Rchb.f.
Disa welwitschii Rchb.f.
Disa welwitschii subsp. **occultans** (Schltr.) H.P.Linder
Disa welwitschii subsp. **welwitschii**
Disa zombica N.E.Br.

MALAYSIA / MALAISIE (LA) / MALASIA

Cymbidium aloifolium (L.) Sw.
Cymbidium atropurpureum (Lindl.) Rolfe
Cymbidium bicolor Lindl.

Cymbidium bicolor subsp. **pubescens** (Lindl.) Du Puy & P.J.Cribb
Cymbidium borneense J.J.Wood
Cymbidium chloranthum Lindl.
Cymbidium dayanum Rchb.f.
Cymbidium elongatum J.J.Wood, Du Puy & Shim
Cymbidium ensifolium (L.) Sw.
Cymbidium ensifolium subsp. **haematodes** (Lindl.) Du Puy & P.J.Cribb
Cymbidium finlaysonianum Lindl.
Cymbidium kinabaluense K.M.Wong & C.L.Chan
Cymbidium lancifolium Hook.
Cymbidium rectum Ridl.
Cymbidium roseum J.J.Sm.
Cymbidium sigmoideum J.J.Sm.
Dendrobium aciculare Lindl.
Dendrobium alabense J.J.Wood
Dendrobium aloifolium (Blume) Rchb.f.
Dendrobium anosmum Lindl.
Dendrobium anthrene Ridl.
Dendrobium aphyllum (Roxb.) Fischer
Dendrobium arcuatum J.J.Sm.
Dendrobium babiense J.J.Sm.
Dendrobium blumei Lindl.
Dendrobium bostrychodes Rchb.f.
Dendrobium calcariferum Carr
Dendrobium calicopis Ridl.
Dendrobium cinereum J.J.Sm.
Dendrobium cinnabarinum Rchb.f.
Dendrobium cinnabarinum var. **angustitepalum** Carr
Dendrobium cinnabarinum var. **cinnabarinum**
Dendrobium cinnabarinum var. **lamelliferum** Carr
Dendrobium clavator Ridl.
Dendrobium compressimentum J.J.Sm.
Dendrobium crabro Ridl.
Dendrobium crocatum Hook.f.
Dendrobium cruentum Rchb.f.
Dendrobium crumenatum Sw.
Dendrobium cumulatum Lindl.
Dendrobium cymboglossum J.J.Wood & A.Lamb
Dendrobium cymbulipes J.J.Sm.
Dendrobium eriiflorum Griff.
Dendrobium erosum (Blume) Lindl.
Dendrobium erythropogon Rchb.f.
Dendrobium farmeri Paxton
Dendrobium fimbriatum Hook.
Dendrobium gracile (Blume) Lindl.
Dendrobium grande Hook.f.
Dendrobium gynoglottis Carr
Dendrobium hamaticalcar J.J.Wood & Dauncey
Dendrobium hasseltii (Blume) Lindl.
Dendrobium hendersonii A.D.Hawkes & A.H.Heller
Dendrobium hercoglossum Rchb.f.

Dendrobium heterocarpum Lindl.
Dendrobium hymenopterum Hook.f.
Dendrobium incurvociliatum J.J.Sm.
Dendrobium incurvum Lindl.
Dendrobium indivisum (Blume) Miq.
Dendrobium indivisum var. **indivisum**
Dendrobium indivisum var. **pallidum** Seidenf.
Dendrobium junceum Lindl.
Dendrobium kiauense Ames & C.Schweinf.
Dendrobium lamellatum (Blume) Lindl.
Dendrobium lamelluliferum J.J.Sm.
Dendrobium lampongense J.J.Sm.
Dendrobium lancilobum J.J.Wood
Dendrobium lawiense J.J.Sm.
Dendrobium leonis (Lindl.) Rchb.f.
Dendrobium limii J.J.Wood
Dendrobium linguella Rchb.f.
Dendrobium lobatum (Blume) Miq.
Dendrobium lobbi Teijsm. & Binnend.
Dendrobium lobulatum Rolfe & J.J.Sm.
Dendrobium lowii Lindl.
Dendrobium lucens Rchb.f.
Dendrobium mannii Ridl.
Dendrobium minimum Ames & C.Schweinf.
Dendrobium modestissimum Kraenzl.
Dendrobium nycteridoglossum Rchb.f.
Dendrobium ovipostoriferum J.J.Sm.
Dendrobium panduriferum Hook.f.
Dendrobium parthenium Rchb.f.
Dendrobium patentilobum Ames & C.Schweinf.
Dendrobium pictum Lindl.
Dendrobium planibulbe Lindl.
Dendrobium pogoniates Rchb.f.
Dendrobium prostratum Ridl.
Dendrobium pseudoaloifolium J.J.Wood
Dendrobium puberilingue J.J.Sm.
Dendrobium pulchellum Roxb. ex Lindl.
Dendrobium radians Rchb.f.
Dendrobium rhodostele Ridl.
Dendrobium rosellum Ridl.
Dendrobium sanderianum Rolfe
Dendrobium sanguinolentum Lindl.
Dendrobium sarawakense Ames
Dendrobium sculptum Rchb.f.
Dendrobium secundum (Blume) Lindl. ex Wall.
Dendrobium setifolium Ridl.
Dendrobium singkawangense J.J.Sm.
Dendrobium smithianum Schltr.
Dendrobium spectatissimum Rchb.f.
Dendrobium spegidoglossum Rchb.f.
Dendrobium stuartii F.M.Bailey

Dendrobium subulatoides Schltr.
Dendrobium swartzii A.D.Hawkes & A.H.Heller
Dendrobium terminale Parish & Rchb.f.
Dendrobium tetrachromum Rchb.f.
Dendrobium tortile Lindl.
Dendrobium tricuspe (Blume) Lindl.
Dendrobium tridentatum Ames & C.Schweinf.
Dendrobium truncatum Lindl.
Dendrobium uncatum Lindl.
Dendrobium ventripes Carr
Dendrobium xanthoacron Schltr.
Dendrobium xiphophyllum Schltr.

MEXICO / MEXIQUE (LE) / MÉXICO

Dracula pusilla (Rolfe) Luer
Encyclia abbreviata (Schltr.) Dressler
Encyclia adenocarpon (La Llave & Lex.) Schltr.
Encyclia adenocaula (La Llave & Lex.) Schltr.
Encyclia adenocaula var. kennedyi (Fowlie & Withner) Hágsater
Encyclia aenicta Dressler & G.E.Pollard
Encyclia alata (Bateman) Schltr.
Encyclia ambigua (Lindl.) Schltr.
Encyclia asperula Dressler & G.E.Pollard
Encyclia atrorubens (Rolfe) Schltr.
Encyclia baculus (Rchb.f.) Dressler & G.E.Pollard
Encyclia bicamerata (Rchb.f.) Dressler & G.E.Pollard
Encyclia boothiana (Lindl.) Dressler
Encyclia boothiana subsp. boothiana
Encyclia boothiana subsp. favoris (Rchb.f.) Dressler & G.E.Pollard
Encyclia brachiata (A.Rich. & H.Galeotti) Dressler & Pollard
Encyclia bractescens (Lindl.) Hoehne
Encyclia brassavolae (Rchb.f.) Dressler
Encyclia campylostalix (Lindl.) Schltr.
Encyclia candollei (Lindl.) Schltr.
Encyclia cepiforme Hooker
Encyclia ceratistes (Lindl.) Schltr.
Encyclia chacaoensis (Rchb.f.) Dressler & G.E.Pollard
Encyclia chiapasensis Withner & Hunt
Encyclia chondylobulbon (A.Rich. & H.Galeotti) Dressler & Pollard
Encyclia citrina (La Llave & Lex.) Dressler
Encyclia cochleata (L.) Lemée
Encyclia concolor (La Llave & Lex.) Schltr.
Encyclia cordigera (Kunth) Dressler
Encyclia cretacea Dressler & G.E.Pollard
Encyclia cyanocolumna (Ames, F.T.Hubb. & C.Schweinf.) Dressler
Encyclia dickinsoniana (Withner) Withner
Encyclia diota (Lindl.) Schltr.
Encyclia distantiflora (A.Rich. & H.Galeotti) Dressler & Pollard
Encyclia fragrans (Sw.) Lemée
Encyclia ghiesbreghtiana (A.Rich. & H.G.Galeotti) Dressler

Encyclia glauca (Knowles & Westc.) Dressler & G.E.Pollard
Encyclia gravida (Lindl.) Schltr.
Encyclia guatemalensis (Klotzsch) Withner
Encyclia hanburii (Lindl.) Schltr.
Encyclia hastata (Lindl.) Dressler & G.E.Pollard
Encyclia incumbens (Lindl.) Mabb.
Encyclia ionophlebium (Rchb.f.) Dressler
Encyclia kennedyi (Fowlie & Withner) Hágsater
Encyclia kienastii (Rchb.f.) Dressler & G.E.Pollard
Encyclia lancifolia (Lindl.) Dressler & G.E.Pollard
Encyclia linkiana (Klotzsch) Schltr.
Encyclia livida (Lindl.) Dressler
Encyclia lorata Dressler & G.E.Pollard
Encyclia luteorosea (A.Rich. & H.Galeotti) Dressler & Pollard
Encyclia maculosa (Ames, F.T.Hubb. & C.Schweinf.) Hoehne
Encyclia magnispatha (Ames, F.T.Hubb. & C.Schweinf.) Dressler
Encyclia mariae (Ames) Hoehne
Encyclia meliosma (Rchb.f.) Schltr.
Encyclia michuacana (La Llave & Lex.) Schltr.
Encyclia microbulbon (Hook.) Schltr.
Encyclia nematocaulon (A.Rich.) Acuña
Encyclia neurosa (Ames) Dressler & Pollard
Encyclia obpyribulbon Hágsater
Encyclia ochracea (Lindl.) Dressler
Encyclia oestlundii (Ames, F.T.Hubb. & Schweinf.) Hágsater & Sterm
Encyclia panthera (Rchb.f.) Schltr.
Encyclia papillosa (Bateman) Aguirre-Olav.
Encyclia parviflora Withner
Encyclia peraltense (Ames) Withner
Encyclia perplexa (Ames, F.T.Hubb. & C.Schweinf.) Withner
Encyclia pollardiana (Withner) Dressler & G.E.Pollard
Encyclia polybulbon (Sw.) Dressler
Encyclia pringlei (Rolfe ex Ames) Schltr.
Encyclia pseudopygmaea (Finet) Dressler & G.E.Pollard
Encyclia pterocarpa (Lindl.) Dressler
Encyclia pygmaea (Hook.) Dressler
Encyclia radiata (Lindl.) Dressler
Encyclia rhombilabia S.Rosillo
Encyclia rhynchophora (A.Rich. & H.G.Galeotti) Dressler
Encyclia selligera (Lindl.) Schltr.
Encyclia semiaperta Hágsater
Encyclia spatella (Rchb.f.) Schltr.
Encyclia suaveolens Dressler
Encyclia subulatifolia (A.Rich. & H.G.Galeotti) Dressler
Encyclia tenuissima (Ames, F.T.Hubb. & C.Schweinf.) Dressler
Encyclia trachycarpa (Lindl. ex Benth) Schltr.
Encyclia tripunctata (Lindl.) Dressler
Encyclia tuerckheimii Schltr.
Encyclia vagans (Ames) Dressler & G.E.Pollard
Encyclia varicosa Bateman ex Lindl.
Encyclia varicosa subsp. *leiobulbon* (Hook.) Dressler & G.E.Pollard

Encyclia varicosa subsp. varicosa
Encyclia venosa (Lindl.) Schltr.
Encyclia vespa (Vell.) Dressler
Encyclia virens (Lindl.) Schltr.
Encyclia vitellina (Lindl.) Dressler

MICRONESIA (FEDERATED STATES OF) / MICRONÉSIA (ETATS FÉDÉRÉS DE) / MICRONESIA (ESTADOS FEDERADOS DE)

Dendrobium delicatulum Kraenzl.
Dendrobium delicatulum subsp. delicatulum
Dendrobium kraemeri Schltr.
Dendrobium violaceominiatum Schltr.

MOZAMBIQUE / MOZAMBIQUE (LE) / MOZAMBIQUE

Disa aconitoides Sond.
Disa aconitoides subsp. aconitoides
Disa aconitoides subsp. concinna (N.E.Br.) H.P.Linder
Disa erubescens Rendle
Disa erubescens subsp. carsonii (N.E.Br.) H.P.Linder
Disa erubescens subsp. erubescens
Disa equestris Rchb.f.
Disa fragrans Schltr.
Disa fragrans subsp. fragrans
Disa hircicornis Rchb.f.
Disa miniata Summerh.
Disa ornithantha Schltr.
Disa saxicola Schltr.
Disa versicolor Rchb.f.
Disa welwitschii Rchb.f.
Disa welwitschii subsp. welwitschii
Disa zimbabweensis H.P.Linder
Disa zombica N.E.Br.

MYANMAR / MYANMAR (LE) / MYANMAR

Cymbidium aloifolium (L.) Sw.
Cymbidium bicolor Lindl.
Cymbidium bicolor subsp. obtusum Du Puy & P.J.Cribb
Cymbidium cochleare Lindl.
Cymbidium cyperifolium Wall. ex Lindl.
Cymbidium cyperifolium subsp. arrogans Du Puy & P.J.Cribb
Cymbidium eburneum Lindl.
Cymbidium elegans Lindl.
Cymbidium erythraeum Lindl.
Cymbidium iridioides D.Don
Cymbidium lancifolium Hook.
Cymbidium lowianum (Rchb.f.) Rchb.f.
Cymbidium lowianum var. iansonii (Rolfe) P.J.Cribb & Du Puy
Cymbidium lowianum var. lowianum

Cymbidium macrorhizon Lindl.
Cymbidium mastersii Griff. ex Lindl.
Cymbidium parishii Rchb.f.
Cymbidium sinense (G.Jacks. & H.C.Andr.) Wild.
Cymbidium suavissimum Hort. ex C.H.Curtis
Cymbidium tigrinum Parish ex Hook.
Cymbidium trachyanum L.Castle
Dendrobium acinaciforme Roxb.
Dendrobium aduncum Wall. ex Lindl.
Dendrobium albosanguineum Lindl.
Dendrobium aloifolium (Blume) Rchb.f.
Dendrobium amoenum Lindl.
Dendrobium anceps Sw.
Dendrobium anosmum Lindl.
Dendrobium aphyllum (Roxb.) Fischer
Dendrobium bellatulum Rolfe
Dendrobium bensoniae Rchb.f.
Dendrobium bicameratum Lindl.
Dendrobium brymerianum Rchb.f.
Dendrobium brymerianum var. **brymerianum**
Dendrobium brymerianum var. **histrionicum** Rchb.f.
Dendrobium capillipes Rchb.f.
Dendrobium cariniferum Rchb.f.
Dendrobium chrysanthum Wall.
Dendrobium chryseum Rolfe
Dendrobium chrysocrepis Parish & Rchb.f.
Dendrobium chrysotoxum Lindl.
Dendrobium compactum Rolfe ex W.Hackett
Dendrobium concinum Miq
Dendrobium crepidatum Lindl. & Paxton
Dendrobium cretaceum Lindl.
Dendrobium cruentum Rchb.f.
Dendrobium crumenatum Sw.
Dendrobium crystallinum Rchb.f.
Dendrobium cumulatum Lindl.
Dendrobium curviflorum Rolfe
Dendrobium cuspidatum Lindl.
Dendrobium delacourii Guill.
Dendrobium densiflorum Wall. ex Lindl.
Dendrobium devonianum Paxton
Dendrobium dickasonii L.O.Williams
Dendrobium dixanthum Rchb.f.
Dendrobium draconis Rchb.f.
Dendrobium eriiflorum Griff.
Dendrobium falconeri Hook.
Dendrobium farmeri Paxton
Dendrobium fimbriatum Hook.
Dendrobium findlayanum Parish & Rchb.f.
Dendrobium formosum Roxb. ex Lindl.
Dendrobium fytchianum Bateman
Dendrobium gibsonii Lindl.

Dendrobium grande Hook.f.
Dendrobium gratiosissimum Rchb.f.
Dendrobium gregulus Seidenf.
Dendrobium griffithianum Lindl.
Dendrobium harveyanum Rchb.f.
Dendrobium heterocarpum Lindl.
Dendrobium incurvum Lindl.
Dendrobium indivisum (Blume) Miq.
Dendrobium indivisum var. **indivisum**
Dendrobium indivisum var. **pallidum** Seidenf.
Dendrobium infundibulum Lindl.
Dendrobium jenkinsii Wall. ex Lindl.
Dendrobium lamellatum (Blume) Lindl.
Dendrobium lasioglossum Rchb.f.
Dendrobium leucochlorum Rchb.f.
Dendrobium lindleyi Steud.
Dendrobium lituiflorum Lindl.
Dendrobium longicornu Wall. ex Lindl.
Dendrobium marmoratum Rchb.f.
Dendrobium moschatum (Buch.-Ham.) Sw.
Dendrobium moulmeinense Rchb.f.
Dendrobium nathanielis Rchb.f.
Dendrobium nobile Lindl.
Dendrobium nobile var. **nobile**
Dendrobium ochreatum Lindl.
Dendrobium pachyglossum Par. & Rchb.f.
Dendrobium palpebrae Lindl.
Dendrobium panduriferum Hook.f.
Dendrobium parcum Rchb.f.
Dendrobium parishii Rchb.f.
Dendrobium pauciflorum King & Pantl.
Dendrobium peguanum Lindl.
Dendrobium pendulum Roxb.
Dendrobium podagraria Hook.f.
Dendrobium porphyrochilum Lindl.
Dendrobium primulinum Lindl.
Dendrobium pulchellum Roxb. ex Lindl.
Dendrobium pychnostachyum Lindl.
Dendrobium rhodopteryguim Rchb.f.
Dendrobium scabrilingue Lindl.
Dendrobium secundum (Blume) Lindl. ex Wall.
Dendrobium senile Parish & Rchb.f.
Dendrobium signatum Rchb.f.
Dendrobium spegidoglossum Rchb.f.
Dendrobium strongylanthum Rchb.f.
Dendrobium stuposum Lindl.
Dendrobium sulcatum Lindl.
Dendrobium sutepense Rolfe ex Downie
Dendrobium terminale Parish & Rchb.f.
Dendrobium thyrsiflorum Rchb.f. ex Andre
Dendrobium tortile Lindl.

Dendrobium transparens Wall. ex Lindl.
Dendrobium trigonopus Rchb.f.
Dendrobium venustum Teijsm. & Binn.
Dendrobium virgineum Rchb.f.
Dendrobium wardianum Warner
Dendrobium wattii (Hook.f.) Rchb.f.
Dendrobium williamsonii Day & Rchb.f.
Dendrobium wilmsianum Schltr.
Dendrobium xanthophlebium Lindl.

NEPAL / NÉPAL / NEPAL

Cymbidium aloifolium (L.) Sw.
Cymbidium bicolor Lindl.
Cymbidium bicolor subsp. **obtusum** Du Puy & P.J.Cribb
Cymbidium cyperifolium Wall. ex Lindl.
Cymbidium cyperifolium subsp. **cyperifolium**
Cymbidium devonianum Paxton
Cymbidium eburneum Lindl.
Cymbidium elegans Lindl.
Cymbidium faberi Rolfe
Cymbidium faberi var. **szechuanicum** (Y.S.Wu & S.C.Chen) Y.S.Wu & S.C.Chen
Cymbidium hookerianum Rchb.f.
Cymbidium iridioides D.Don
Cymbidium lancifolium Hook.
Cymbidium macrorhizon Lindl.
Dendrobium amoenum Lindl.
Dendrobium anceps Sw.
Dendrobium aphyllum (Roxb.) Fischer
Dendrobium bicameratum Lindl.
Dendrobium capillipes Rchb.f.
Dendrobium chrysanthum Wall.
Dendrobium chryseum Rolfe
Dendrobium chrysotoxum Lindl.
Dendrobium crepidatum Lindl. & Paxton
Dendrobium cretaceum Lindl.
Dendrobium cumulatum Lindl.
Dendrobium densiflorum Wall. ex Lindl.
Dendrobium denudans D.Don
Dendrobium eriiflorum Griff.
Dendrobium farmeri Paxton
Dendrobium fimbriatum Hook.
Dendrobium formosum Roxb. ex Lindl.
Dendrobium gibsonii Lindl.
Dendrobium griffithianum Lindl.
Dendrobium heterocarpum Lindl.
Dendrobium longicornu Wall. ex Lindl.
Dendrobium monticola Hunt & Summerh.
Dendrobium moschatum (Buch.-Ham.) Sw.
Dendrobium nobile Lindl.
Dendrobium nobile var. **nobile**

Dendrobium peguanum Lindl.
Dendrobium porphyrochilum Lindl.
Dendrobium primulinum Lindl.
Dendrobium pulchellum Roxb. ex Lindl.
Dendrobium tortile Lindl.
Dendrobium transparens Wall. ex Lindl.

NEW CALEDONIA

Dendrobium bowmanii Benth.
Dendrobium comptonii Rendle
Dendrobium macranthum A.Rich.
Dendrobium macropus (Endl.) Rchb.f. ex Lindl.
Dendrobium masarangense Schltr
Dendrobium masarangense subsp. masarangense
Dendrobium sylvanum Rchb.f.

NEW ZEALAND / NOUVELLE-ZÉLANDE (LA) / NUEVA ZELANDIA

Dendrobium cunninghamii Lindl.

NICARAGUA / NICARAGUIA (LE) / NICARAGUA

Dracula pusilla (Rolfe) Luer
Dracula vespertilio (Rchb.f.) Luer
Encyclia abbreviata (Schltr.) Dressler
Encyclia adenocarpon (La Llave & Lex.) Schltr.
Encyclia aenicta Dressler & G.E.Pollard
Encyclia alata (Bateman) Schltr.
Encyclia amanda (Ames) Dressler
Encyclia ambigua (Lindl.) Schltr.
Encyclia baculus (Rchb.f.) Dressler & G.E.Pollard
Encyclia ceratistes (Lindl.) Schltr.
Encyclia chacaoensis (Rchb.f.) Dressler & G.E.Pollard
Encyclia dickinsoniana (Withner) Withner
Encyclia luteorosea (A.Rich. & H.Galeotti) Dressler & Pollard
Encyclia mooreana (Rolfe) Schltr.
Encyclia ochracea (Lindl.) Dressler
Encyclia polybulbon (Sw.) Dressler
Encyclia pygmaea (Hook.) Dressler
Encyclia rhynchophora (A.Rich. & H.G.Galeotti) Dressler
Encyclia selligera (Lindl.) Schltr.
Encyclia stellata (Lindl.) Schltr.
Encyclia vagans (Ames) Dressler & G.E.Pollard
Encyclia vespa (Vell.) Dressler
Encyclia virens (Lindl.) Schltr.

NIGERIA / NIGÉRIA (LE) / NIGERIA

Disa equestris Rchb.f.
Disa erubescens Rendle
Disa erubescens subsp. erubescens

Disa hircicornis Rchb.f.
Disa nigerica Rolfe
Disa ochrostachya Rchb.f.
Disa scutellifera A.Rich.
Disa welwitschii Rchb.f.
Disa welwitschii subsp. **occultans** (Schltr.) H.P.Linder
Disa welwitschii subsp. **welwitschii**

PAKISTAN / PAKISTAN (LE) / PAKISTÁN

Cymbidium macrorhizon Lindl.

PALAU / PALAOS / PALAU

Dendrobium pachystele Schltr.

PANAMA / PANAMA (LE) / PANAMÁ

Dracula erythrochaete (Rchb.f.) Luer
Dracula pusilla (Rolfe) Luer
Encyclia abbreviata (Schltr.) Dressler
Encyclia amanda (Ames) Dressler
Encyclia brassavolae (Rchb.f.) Dressler
Encyclia campylostalix (Lindl.) Schltr.
Encyclia candollei (Lindl.) Schltr.
Encyclia ceratistes (Lindl.) Schltr.
Encyclia chacaoensis (Rchb.f.) Dressler & G.E.Pollard
Encyclia chimborazoensis (Schltr.) Dressler
Encyclia cochleata (L.) Lemée
Encyclia cordigera (Kunth) Dressler
Encyclia fortunae Dressler
Encyclia livida (Lindl.) Dressler
Encyclia mooreana (Rolfe) Schltr.
Encyclia prismatocarpa (Rchb.f.) Dressler
Encyclia pseudopygmaea (Finet) Dressler & G.E.Pollard
Encyclia pygmaea (Hook.) Dressler
Encyclia sima Dressler
Encyclia spondiadum (Rchb.f.) Dressler
Encyclia stellata (Lindl.) Schltr.
Encyclia varicosa Bateman ex Lindl.
Encyclia varicosa subsp. **varicosa**
Encyclia vespa (Vell.) Dressler

PAPUA NEW GUINEA / PAPOUASIE-NOUVELLE-GUINÉE (LA) / PAPUA NUEVA GUINEA

Cymbidium ensifolium (L.) Sw.
Cymbidium ensifolium subsp. **haematodes** (Lindl.) Du Puy & P.J.Cribb
Cymbidium lancifolium Hook.
Dendrobium aberrans Schltr.
Dendrobium acutisepalum J.J.Sm.
Dendrobium alaticaulinum P.Royen

Dendrobium alexandrae Schltr.
Dendrobium amphigenyum Ridl.
Dendrobium andreemillarae T.M.Reeve
Dendrobium anosmum Lindl.
Dendrobium antennatum Lindl.
Dendrobium apertum Schltr.
Dendrobium arachnoglossum Rchb.f.
Dendrobium armeniacum P.J.Cribb
Dendrobium atroviolaceum Rolfe
Dendrobium aurantiroseum P.Royen ex T.M.Reeve
Dendrobium baeuerlenii F.Muell. & Kraenzl.
Dendrobium bifalce Lindl.
Dendrobium bigibbum Lindl.
Dendrobium brachycalyptra Schltr.
Dendrobium bracteosum Rchb.f.
Dendrobium brassii T.M.Reeve & P.Woods
Dendrobium brevicaule Rolfe
Dendrobium brevicaule subsp. calcarium (J.J.Sm.) T.M.Reeve & P.Woods
Dendrobium brevicaule subsp. pentagonum (Kraenzl.) T.M.Reeve
Dendrobium brevilabium Schltr
Dendrobium buffumii A.D.Hawkes
Dendrobium calcaratum A.Rich.
Dendrobium calcaratum subsp. papillatum Dauncey
Dendrobium caliculimentum R.S.Rogers
Dendrobium canaliculatum R.Br.
Dendrobium canaliculatum var. canaliculatum
Dendrobium capituliflorum Rolfe
Dendrobium carronii Lavarack & P.J.Cribb
Dendrobium chordiforme Kraenzl.
Dendrobium cochleatum J.J.Sm.
Dendrobium cochliodes Schltr.
Dendrobium codonosepalum J.J.Sm.
Dendrobium conanthum Schltr.
Dendrobium constrictum J.J.Sm.
Dendrobium convolutum Rolfe
Dendrobium crispilinguum P.J.Cribb
Dendrobium cruttwellii T.M.Reeve
Dendrobium cuthbertsonii F.Muell.
Dendrobium cyanocentrum Schltr.
Dendrobium dekockii J.J.Sm.
Dendrobium delicatulum Kraenzl.
Dendrobium delicatulum subsp. delicatulum
Dendrobium delicatulum subsp. huliorum T.M.Reeve & P.Woods
Dendrobium delphinioides R.Rogers
Dendrobium dendrocolloides J.J.Sm.
Dendrobium diceras Schltr.
Dendrobium dichaeoides Schltr.
Dendrobium dichroma Schltr.
Dendrobium dillonianum A.D.Hawkes & A.H.Heller
Dendrobium discolor Lindl.
Dendrobium discolor var. discolor

Dendrobium eboracense Kraenzl.
Dendrobium engae T.M.Reeve
Dendrobium euryanthum Schltr.
Dendrobium eumelinum Schltr.
Dendrobium eximium Schltr.
Dendrobium finisterrae Schltr.
Dendrobium flagellum Schltr.
Dendrobium flammula Schltr.
Dendrobium forbesii Ridl.
Dendrobium fractum T.M.Reeve
Dendrobium fulgidum Bateman ex Lindl.
Dendrobium fulgidum var. fulgidum
Dendrobium geotropum T.M.Reeve
Dendrobium goldfinchii F.Muell.
Dendrobium gouldii Rchb.f.
Dendrobium habbemense Van Royen
Dendrobium hamiferum P.J.Cribb
Dendrobium helix P.J.Cribb
Dendrobium hellwigianum Kraenzl.
Dendrobium hodgkinsonii Rolfe
Dendrobium hymenocentrum Schltr.
Dendrobium inamoenum Kraenzl.
Dendrobium × vonpaulsenianum A.Hawkes
Dendrobium johannis Rchb.f.
Dendrobium johnsoniae F.Muell.
Dendrobium juncoideum P. Royen
Dendrobium kauldorumii T.M.Reeve
Dendrobium laevifolium Stapf
Dendrobium lanepoolei R.Rogers
Dendrobium lasianthera J.J.Sm.
Dendrobium lawesii F.Muell.
Dendrobium leucocyanum T.M.Reeve
Dendrobium leucohybos Schltr.
Dendrobium lineale Rolfe
Dendrobium litorale Schltr.
Dendrobium lobii Teijsm. & Binnend.
Dendrobium loesenerianum Schltr.
Dendrobium macrogenion Schltr.
Dendrobium macrophyllum A.Rich.
Dendrobium macrophyllum var. macrophyllum
Dendrobium macrophyllum var. subvelutinum J.J.Sm.
Dendrobium magistratus P.J.Cribb
Dendrobium masarangense Schltr
Dendrobium masarangense subsp. masarangense
Dendrobium masarangense var. theionanthum (Schltr.) T.M.Reeve & P.Woods
Dendrobium mayandyi T.M.Reeve & Renz
Dendrobium melinanthum Schltr.
Dendrobium mirbelianum Gaudich.
Dendrobium molle J.J.Sm.
Dendrobium montis-yulei Kraenzl.
Dendrobium mystroglossum Schltr.

Dendrobium nardoides Schltr.
Dendrobium navicula Kraenzl.
Dendrobium nebularum Schltr.
Dendrobium nindii W.Hill
Dendrobium nothofagicola T.M.Reeve
Dendrobium nubigenum Schltr.
Dendrobium obtusum J.J.Sm.
Dendrobium oreodoxa Schltr.
Dendrobium oreogenum Schltr.
Dendrobium otaguroanum A.D.Hawkes
Dendrobium pachystele Schltr.
Dendrobium pachythrix T.M.Reeve & P.Woods
Dendrobium pentapterum Schltr.
Dendrobium petiolatum Schltr.
Dendrobium pleurodes Schltr.
Dendrobium polysema Schltr.
Dendrobium praetermissum Dauncey
Dendrobium pseudocalceolum J.J.Sm.
Dendrobium pseudoglomeratum T.M.Reeve & J.J.Woods
Dendrobium pseudopeloricum J.J.Sm.
Dendrobium punamense Schltr
Dendrobium puniceum Ridl.
Dendrobium putnamii Hawkes & Heller
Dendrobium rarum Schltr.
Dendrobium rarum var. miscegeneum Dauncey
Dendrobium rarum var. pelorium Dauncey
Dendrobium rarum var. rarum
Dendrobium rhabdoglossum Schltr.
Dendrobium rhodostictum F.Muell. & Kraenzl.
Dendrobium rigidifolium Rolfe
Dendrobium rigidum R.Br.
Dendrobium roseicolor A.D.Hawkes & A.H.Heller
Dendrobium roseipes Schltr.
Dendrobium roseum Schltr.
Dendrobium rupestre J.J.Sm.
Dendrobium salmoneum Schltr.
Dendrobium × schumannianum Schltr.
Dendrobium simplex J.J.Sm.
Dendrobium smillieae F.Muell.
Dendrobium soriense Howcroft
Dendrobium speciosum Sm.
Dendrobium spectabile (Blume) Miq.
Dendrobium stolleanum Schltr.
Dendrobium stuartii F.M.Bailey
Dendrobium subacaule Reinw. ex Lindl.
Dendrobium subclausum Rolfe
Dendrobium subclausum var. pandanicola J.J.Wood
Dendrobium subclausum var. phlox (Schltr.) J.J.Wood
Dendrobium subclausum var. speciosum J.J.Wood
Dendrobium subclausum var. subclausum
Dendrobium subquadratum J.J.Sm.

Dendrobium subuliferum J.J.Sm.
Dendrobium sulphureum Schltr.
Dendrobium sulphureum var. **cellulosum** (J.J.Sm.) T.M.Reeve & P.Woods
Dendrobium sulphureum var. **rigidifolium** T.M.Reeve & P.Woods
Dendrobium sulphureum var. **sulphureum**
Dendrobium sylvanum Rchb.f.
Dendrobium tangerinum P.J.Cribb
Dendrobium tapiniense T.M.Reeve
Dendrobium terrestre J.J.Sm.
Dendrobium torricellense Schltr.
Dendrobium trachythece Schltr.
Dendrobium trilamellatum J.J.Sm.
Dendrobium undatialatum Schltr.
Dendrobium vannouhuysii J.J.Sm.
Dendrobium verruculosum Schltr.
Dendrobium vexillarius J.J.Sm.
Dendrobium vexillarius var. **albiviride** (Van Royen) T.M.Reeve & P.Woods
Dendrobium vexillarius var. **elworthyi** T.M.Reeve & P.Woods
Dendrobium vexillarius var. **microblepharum** (Schltr.) T.M.Reeve & P.Woods
Dendrobium vexillarius var. **retroflexum** (J.J.Sm.) T.M.Reeve & P.Woods
Dendrobium vexillarius var. **uncinatum** (Schltr.) T.M.Reeve & P.Woods
Dendrobium vexillarius var. **vexillarius**
Dendrobium violaceum Kraenzl.
Dendrobium violaceum subsp. **cyperifolium** (Schltr.) T.M.Reeve & P.Woods
Dendrobium violaceum subsp. **violaceum**
Dendrobium wentianum J.J.Sm.
Dendrobium williamsianum Rchb.f.
Dendrobium womersleyi T.M.Reeve
Dendrobium womersleyi var. **autophilum** Dauncey
Dendrobium womersleyi var. **womersleyi**
Dendrobium woodsii P.J.Cribb
Dendrobium wulaiense Howcroft
Dendrobium xanthogenium Schltr.
Dendrobium × yengiliense T.M.Reeve

PARAGUAY / PARAGUAY (LE) / PARAGUAY (EL)

Encyclia inversa (Lindl.) Pabst
Encyclia linearifolioides (Kraenzl.) Hoehne

PERU / PÉROU / PERÚ

Dracula janetiae (Luer) Luer
Encyclia aemula (Lindl.) Carnevali & Ramírez
Encyclia chloroleuca (Hook.) Neum.
Encyclia cordigera (Kunth.) Dressler
Encyclia cyperifolia (C. Schweinf.) Carnevali & Ramírez
Encyclia grammatoglossa (Rchb.f.) Dressler
Encyclia hartwegii (Lindl.) R.Vásquez & Dodson
Encyclia livida (Lindl.) Dressler
Encyclia luteorosea (A.Rich. & H.Galeotti) Dressler & Pollard

Encyclia microtos (Rchb.f.) Hoehne
Encyclia pulcherrima Dodson & Bennett
Encyclia sclerocladia (Lindl. ex Rchb.f.) Hoehne
Encyclia vespa (Vell.) Dressler

PHILIPPINES (THE) / PHILIPPINES (LES) / FILIPINAS

Cymbidium atropurpureum (Lindl.) Rolfe
Cymbidium bicolor Lindl.
Cymbidium bicolor subsp. pubescens (Lindl.) Du Puy & P.J.Cribb
Cymbidium cyperifolium Wall. ex Lindl.
Cymbidium cyperifolium subsp. arrogans Du Puy & P.J.Cribb
Cymbidium dayanum Rchb.f.
Cymbidium ensifolium (L.) Sw.
Cymbidium ensifolium subsp. ensifolium
Cymbidium finlaysonianum Lindl.
Dendrobium albayense Ames
Dendrobium amethystoglossum Rchb.f.
Dendrobium anosmum Lindl.
Dendrobium auriculatum Ames & Quisumb.
Dendrobium basilanense Ames
Dendrobium blumei Lindl.
Dendrobium bullenianum Rchb.f.
Dendrobium ceraula Rchb.f.
Dendrobium chameleon Ames
Dendrobium crumenatum Sw.
Dendrobium cumulatum Lindl.
Dendrobium dearei Rchb.f.
Dendrobium escritorii Ames
Dendrobium goldschmidtianum Kraenzl.
Dendrobium heterocarpum Lindl.
Dendrobium junceum Lindl.
Dendrobium linguella Rchb.f.
Dendrobium lunatum Lindl.
Dendrobium macgregorii F.Muell. ex Kraenzl
Dendrobium macrophyllum A.Rich.
Dendrobium macrophyllum var. macrophyllum
Dendrobium modestum Rchb.f.
Dendrobium multiramosum Ames
Dendrobium obrienianum Kraenzl.
Dendrobium papilio Loher
Dendrobium pristinum Ames
Dendrobium profusum Rchb.f.
Dendrobium ramosii Ames
Dendrobium sanderae Rolfe
Dendrobium sanguinolentum Lindl.
Dendrobium schroederi Rolfe
Dendrobium schuetzei Rolfe
Dendrobium secundum (Blume) Lindl. ex Wall.
Dendrobium sinuosum Ames
Dendrobium taurinum Lindl.

Dendrobium usterioides Ames
Dendrobium ventricosum Kraenzl.
Dendrobium victoriae-reginae Loher
Dendrobium wenzelii Ames

REUNION / RÉUNION / REUNIÓN

Disa borbonica Balf.f. & S.Moore

RWANDA / RWANDA (LE) / RWANDA

Disa aconitoides Sond.
Disa aconitoides subsp. **aconitoides**
Disa eminii Kraenzl.
Disa erubescens Rendle
Disa erubescens subsp. **erubescens**
Disa fragrans Schltr.
Disa fragrans subsp. **deckenii** (Rchb.f.) H.P.Linder
Disa fragrans subsp. **fragrans**
Disa hircicornis Rchb.f.
Disa ochrostachya Rchb.f.
Disa robusta N.E.Br.
Disa stairsii Kraenzl.

SAMOA / SAMOA (LE) / SAMOA

Dendrobium calcaratum A.Rich
Dendrobium calcaratum subsp. **calcaratum**
Dendrobium macrophyllum A.Rich.
Dendrobium macrophyllum var. **macrophyllum**
Dendrobium macropus (Endl.) Rchb.f. ex Lindl.
Dendrobium mohlianum Rchb.f.
Dendrobium samoense P.J.Cribb
Dendrobium whistleri P.J.Cribb

SEYCHELLES / SEYCHELLES (LES) / SEYCHELLES

Dendrobium crumenatum Sw.

SINGAPORE / SINGAPOUR / SINGAPUR

Dendrobium aciculare Lindl.
Dendrobium aloifolium (Blume) Rchb.f.
Dendrobium crumenatum Sw.
Dendrobium prostratum Ridl.
Dendrobium secundum (Blume) Lindl. ex Wall.
Dendrobium setifolium Ridl.
Dendrobium spegidoglossum Rchb.f.

SOCIETY ISLANDS (FRENCH)

Dendrobium vagans Schltr.

SOLOMON ISLANDS / ILES SALOMON / ISLAS SALOMÓN

Dendrobium antennatum Lindl.
Dendrobium bifalce Lindl.
Dendrobium calcaratum A.Rich.
Dendrobium calcaratum subsp. **calcaratum**
Dendrobium caliculimentum R.S.Rogers
Dendrobium capituliflorum Rolfe
Dendrobium conanthum Schltr.
Dendrobium delicatulum Kraenzl.
Dendrobium delicatulum subsp. **delicatulum**
Dendrobium erosum (Blume) Lindl.
Dendrobium gnomus Ames
Dendrobium goldfinchii F.Muell.
Dendrobium gouldii Rchb.f.
Dendrobium johnsoniae F.Muell.
Dendrobium laevifolium Stapf
Dendrobium lawesii F.Muell.
Dendrobium macranthum A.Rich.
Dendrobium macrophyllum A.Rich.
Dendrobium macrophyllum var. **macrophyllum**
Dendrobium magistratus P.J.Cribb
Dendrobium masarangense Schltr
Dendrobium masarangense subsp. **masarangense**
Dendrobium mirbelianum Gaudich.
Dendrobium mohlianum Rchb.f.
Dendrobium pachystele Schltr.
Dendrobium petiolatum Schltr.
Dendrobium polysema Schltr.
Dendrobium punamense Schltr.
Dendrobium puniceum Ridl.
Dendrobium rennellii P.J.Cribb
Dendrobium rhodostictum F.Muell. & Kraenzl.
Dendrobium ruginosum Ames
Dendrobium sancristobalense P.J.Cribb
Dendrobium spectabile (Blume) Miq.
Dendrobium subacaule Reinw. ex Lindl.
Dendrobium sylvanum Rchb.f.
Dendrobium undatialatum Schltr.
Dendrobium vagans Schltr.
Dendrobium violaceominiatum Schltr.
Dendrobium whistleri P.J.Cribb

SOUTH AFRICA / AFRIQUE DU SUD (L') / SUDAFRICA

Disa aconitoides Sond.
Disa aconitoides subsp. **aconitoides**
Disa alticola H.P.Linder
Disa amoena H.P.Linder
Disa arida Vlok

Disa aristata H.P.Linder
Disa atricapilla (Harv. ex Lindl.) Bolus
Disa aurata (Bolus) Koopowitz & L.T.Parker
Disa basutorum Schltr.
Disa begleyi L.Bolus
Disa bivalvata (L.f.) T.Durand & Schinz
Disa bodkinii Bolus
Disa brachyceras Lindl.
Disa brevipetala H.P.Linder
Disa caffra Bolus
Disa cardinalis H.P.Linder
Disa caulescens Lindl.
Disa cedarbergensis H.P.Linder
Disa cephalotes Rchb.f.
Disa cephalotes subsp. **cephalotes**
Disa cephalotes subsp. **frigida** (Schltr.) H.P.Linder
Disa chrysostachya Sw.
Disa clavicornis H.P.Linder
Disa cochlearis Johnson & Liltved
Disa cooperi Rchb.f.
Disa cornuta (L.) Sw.
Disa crassicornis Lindl.
Disa cylindrica (Thunb.) Sw.
Disa dracomontana Schelpe ex H.P.Linder
Disa draconis (L.f.) Sw.
Disa elegans Sond. ex Rchb.f.
Disa esterhuyseniae Schelpe ex H.P.Linder
Disa extinctoria Rchb.f.
Disa fasciata Lindl.
Disa ferruginea (Thunb.) Sw.
Disa filicornis (L.f.) Thunb.
Disa fragrans Schltr.
Disa fragrans subsp. **fragrans**
Disa galpinii Rolfe
Disa gladioliflora Burch. ex Lindl.
Disa gladioliflora subsp. **capricornis** (Rchb.f.) H.P.Linder
Disa gladioliflora subsp. **gladioliflora**
Disa glandulosa Burch. ex Lindl.
Disa hallackii Rolfe
Disa harveiana Johnson & Linder
Disa harveiana subsp. **harveiana**
Disa harveiana subsp. **longicalcarata** Johnson & Linder
Disa hircicornis Rchb.f.
Disa introrsa Kurzweil, Liltved & Linder
Disa karooica Johnson & Linder
Disa lineata Bolus
Disa longicornu L.f.
Disa longifolia Lindl.
Disa maculata L.f.
Disa maculomarronina McMurtry
Disa marlothii Bolus

Disa micropetala Schltr.
Disa minor (Sond.) Rchb.f.
Disa montana Sond.
Disa neglecta Sond.
Disa nervosa Lindl.
Disa nivea H.P.Linder
Disa obtusa Lindl.
Disa obtusa subsp. **hottentotica** H.P.Linder
Disa obtusa subsp. **obtusa**
Disa obtusa subsp. **picta** (Sond.) H.P.Linder
Disa ocellata Bolus
Disa oligantha Rchb.f.
Disa oreophila Bolus
Disa oreophila subsp. **erecta** H.P.Linder
Disa oreophila subsp. **oreophila**
Disa ovalifolia Sond.
Disa patula Sond.
Disa patula var. **patula**
Disa patula var. **transvaalensis** Summerh.
Disa pillansii L.Bolus
Disa polygonoides Lindl.
Disa porrecta Sw.
Disa pulchra Sond.
Disa racemosa L.f.
Disa rhodantha Schltr.
Disa richardiana Lehm. ex Bolus
Disa rosea Lindl.
Disa sagittalis (L.f.) Sw.
Disa salteri G.J.Lewis
Disa sanguinea Sond.
Disa sankeyi Rolfe
Disa saxicola Schltr.
Disa schizodioides Sond.
Disa scullyi Bolus
Disa similis Summerh.
Disa stachyoides Rchb.f.
Disa stricta Sond.
Disa subtenuicornis H.P.Linder
Disa telipogonis Rchb.f.
Disa tenella (L.f.) Sw.
Disa tenella subsp. **pusilla** H.P.Linder
Disa tenella subsp. **tenella**
Disa tenuicornis Bolus
Disa tenuifolia Sw.
Disa tenuis Lindl.
Disa thodei Schltr. ex Kraenzl.
Disa triloba Lindl.
Disa tripetaloides (L.f.) N.E.Br.
Disa tysonii Bolus
Disa uncinata Bolus
Disa uniflora P.J.Bergius

Disa vaginata Harv. ex Lindl.
Disa vasselotii Bolus ex Schltr.
Disa venosa Sw.
Disa versicolor Rchb.f.
Disa welwitschii Rchb.f.
Disa welwitschii subsp. **welwitschii**
Disa woodii Schltr.
Disa zimbabweensis H.P.Linder
Disa zuluensis Rolfe

SRI LANKA / SRI LANKA / SRI LANKA

Cymbidium aloifolium (L.) Sw.
Cymbidium bicolor Lindl.
Cymbidium bicolor subsp. **bicolor**
Cymbidium ensifolium (L.) Sw.
Cymbidium ensifolium subsp. **haematodes** (Lindl.) Du Puy & P.J.Cribb
Dendrobium anosmum Lindl.
Dendrobium crumenatum Sw.
Dendrobium heterocarpum Lindl.
Dendrobium macarthiae Thw.
Dendrobium macrostachyum Lindl.

SUDAN / SOUDAN (LE) / SUDÁN (EL)

Disa erubescens Rendle
Disa erubescens subsp. **erubescens**
Disa fragrans Schltr.
Disa fragrans subsp. **deckenii** (Rchb.f.) H.P.Linder
Disa fragrans subsp. **fragrans**
Disa hircicornis Rchb.f.
Disa scutellifera A.Rich.

SURINAM / SURINAME / SURINAME

Encyclia calamaria (Lindl.) Pabst
Encyclia chloroleuca (Hook.) Neum.
Encyclia granitica (Lindl.) Schltr.

SWAZILAND / SWAZILAND (LE) / SWAZILANDIA

Disa chrysostachya Sw.
Disa extinctoria Rchb.f.
Disa intermedia H.P.Linder
Disa nervosa Lindl.
Disa patula Sond.
Disa patula var. **patula**
Disa patula var. **transvaalensis** Summerh.
Disa polygonoides Lindl.
Disa saxicola Schltr.
Disa stachyoides Rchb.f.

Disa versicolor Rchb.f.
Disa woodii Schltr.

TANZANIA, (UNITED REPUBLIC OF) / RÉPUBLIQUE-UNIE DE TANZANIE (LA) / REPÚBLICA UNIDA DE TANZANÍA (LA)

Disa aconitoides Sond.
Disa aconitoides subsp. **aconitoides**
Disa aconitoides subsp. **goetzeana** (Kraenzl.) H.P.Linder
Disa aequiloba Summerh.
Disa aperta N.E.Br.
Disa celata Summerh.
Disa cryptantha Summerh.
Disa eminii Kraenzl.
Disa englerana Kraenzl.
Disa equestris Rchb.f.
Disa erubescens Rendle
Disa erubescens subsp. **carsonii** (N.E.Br.) H.P.Linder
Disa erubescens subsp. **erubescens**
Disa fragrans Schltr.
Disa fragrans subsp. **deckenii** (Rchb.f.) H.P.Linder
Disa fragrans subsp. **fragrans**
Disa hircicornis Rchb.f.
Disa miniata Summerh.
Disa nyikensis H.P.Linder
Disa ochrostachya Rchb.f.
Disa ornithantha Schltr.
Disa perplexa H.P.Linder
Disa robusta N.E.Br.
Disa rungweensis Schltr.
Disa satyriopsis Kraenzl.
Disa saxicola Schltr.
Disa stairsii Kraenzl.
Disa ukingensis Schltr.
Disa walleri Rchb.f.
Disa welwitschii Rchb.f.
Disa welwitschii subsp. **occultans** (Schltr.) H.P.Linder
Disa welwitschii subsp. **welwitschii**
Disa zombica N.E.Br.

THAILAND / THAÏLANDE (LA) / TAILANDIA

Cymbidium aloifolium (L.) Sw.
Cymbidium atropurpureum (Lindl.) Rolfe
Cymbidium bicolor Lindl.
Cymbidium bicolor subsp. **obtusum** Du Puy & P.J.Cribb
Cymbidium cochleare Lindl.
Cymbidium cyperifolium Wall. ex Lindl.
Cymbidium cyperifolium subsp. **arrogans** Du Puy & P.J.Cribb
Cymbidium dayanum Rchb.f.
Cymbidium devonianum Paxton

Cymbidium ensifolium (L.) Sw.
Cymbidium ensifolium subsp. ensifolium
Cymbidium ensifolium subsp. haematodes (Lindl.) Du Puy & P.J.Cribb
Cymbidium erythraeum Lindl.
Cymbidium finlaysonianum Lindl.
Cymbidium insigne Rolfe
Cymbidium lancifolium Hook.
Cymbidium lowianum (Rchb.f.) Rchb.f.
Cymbidium lowianum var. lowianum
Cymbidium macrorhizon Lindl.
Cymbidium mastersii Griff. ex Lindl.
Cymbidium pulchellum L.Castle
Cymbidium sinense (G.Jacks. & H.C.Andr.) De Wild.
Cymbidium trachyanum L.Castle
Dendrobium aciculare Lindl.
Dendrobium acinaciforme Roxb.
Dendrobium aduncum Wall. ex Lindl.
Dendrobium albosanguineum Lindl.
Dendrobium aloifolium (Blume) Rchb.f.
Dendrobium anceps Sw.
Dendrobium anosmum Lindl.
Dendrobium aphyllum (Roxb.) Fischer
Dendrobium bellatulum Rolfe
Dendrobium bensoniae Rchb.f.
Dendrobium bicameratum Lindl.
Dendrobium bilobulatum Seidenf.
Dendrobium blumei Lindl.
Dendrobium brevimentum Seidenf.
Dendrobium brymerianum Rchb.f.
Dendrobium brymerianum var. brymerianum
Dendrobium brymerianum var. histrionicum Rchb.f.
Dendrobium calicopis Ridl.
Dendrobium capillipes Rchb.f.
Dendrobium cariniferum Rchb.f.
Dendrobium christyanum Rchb.f.
Dendrobium chrysanthum Wall.
Dendrobium chryseum Rolfe
Dendrobium chrysotoxum Lindl.
Dendrobium ciliatilabellum Seidenf.
Dendrobium clavator Ridl.
Dendrobium compactum Rolfe ex W.Hackett
Dendrobium confinale Kerr
Dendrobium crepidatum Lindl. & Paxton
Dendrobium cretaceum Lindl.
Dendrobium crocatum Hook.f.
Dendrobium cruentum Rchb.f.
Dendrobium crumenatum Sw.
Dendrobium crystallinum Rchb.f.
Dendrobium cumulatum Lindl.
Dendrobium curviflorum Rolfe
Dendrobium cuspidatum Lindl.

Dendrobium dantaniense Guill.
Dendrobium delacourii Guill.
Dendrobium deltatum Seidenf.
Dendrobium densiflorum Wall. ex Lindl.
Dendrobium devonianum Paxton
Dendrobium dickasonii L.O.Williams
Dendrobium dixanthum Rchb.f.
Dendrobium dixonianum Rolfe ex Downie
Dendrobium draconis Rchb.f.
Dendrobium eriiflorum Griff.
Dendrobium erostelle Seidenf.
Dendrobium erosum (Blume) Lindl.
Dendrobium eserre Seidenf.
Dendrobium exile Schltr.
Dendrobium falconeri Hook.
Dendrobium farmeri Paxton
Dendrobium fesselianum M.Wolff
Dendrobium fimbriatum Hook.
Dendrobium findlayanum Parish & Rchb.f.
Dendrobium formosum Roxb. ex Lindl.
Dendrobium friedericksianum Rchb.f.
Dendrobium fuerstenbergianum Schltr.
Dendrobium garrettii Seidenf.
Dendrobium gibsonii Lindl.
Dendrobium gracile (Blume) Lindl.
Dendrobium grande Hook.f.
Dendrobium gratiosissimum Rchb.f.
Dendrobium gregulus Seidenf.
Dendrobium griffithianum Lindl.
Dendrobium hainanense Rolfe
Dendrobium harveyanum Rchb.f.
Dendrobium hendersonii A.D.Hawkes & A.H.Heller
Dendrobium henryi Schltr.
Dendrobium hercoglossum Rchb.f.
Dendrobium heterocarpum Lindl.
Dendrobium hymenopterum Hook.f.
Dendrobium incurvum Lindl.
Dendrobium indivisum (Blume) Miq.
Dendrobium indivisum var. **indivisum**
Dendrobium indivisum var. **pallidum** Seidenf.
Dendrobium infundibulum Lindl.
Dendrobium intricatum Gagnep.
Dendrobium jenkinsii Wall. ex Lindl.
Dendrobium kanburiense Seidenf.
Dendrobium keithii Ridl.
Dendrobium kratense Kerr
Dendrobium lamellatum (Blume) Lindl.
Dendrobium leonis (Lindl.) Rchb.f.
Dendrobium lindleyi Steud.
Dendrobium linguella Rchb.f.
Dendrobium lituiflorum Lindl.

Dendrobium lobii Teijsm. & Binnend.
Dendrobium lueckelianum Fessel & M.Wolff
Dendrobium mannii Ridl.
Dendrobium monticola Hunt & Summerh.
Dendrobium moschatum (Buch.-Ham.) Sw.
Dendrobium mucronatum Seidenf.
Dendrobium nathanielis Rchb.f.
Dendrobium nobile Lindl.
Dendrobium nobile var. nobile
Dendrobium ochreatum Lindl.
Dendrobium oligophyllum Gagnep.
Dendrobium pacyglossum Pay. & Rchb.f.
Dendrobium palpebrae Lindl.
Dendrobium panduriferum Hook.f.
Dendrobium parciflorum Rchb.f. ex Lindl.
Dendrobium parcum Rchb.f.
Dendrobium parishii Rchb.f.
Dendrobium pauciflorum King & Pantl.
Dendrobium pendulum Roxb.
Dendrobium planibulbe Lindl.
Dendrobium podagraria Hook.f.
Dendrobium porphyrophyllum Guill.
Dendrobium primulinum Lindl.
Dendrobium proteranthum Seidenf.
Dendrobium pulchellum Roxb. ex Lindl.
Dendrobium pychnostachyum Lindl.
Dendrobium rhodostele Ridl.
Dendrobium sanguinolentum Lindl.
Dendrobium salaccence (Blume) Lindl.
Dendrobium scabrilingue Lindl.
Dendrobium secundum (Blume) Lindl. ex Wall.
Dendrobium senile Parish & Rchb.f.
Dendrobium setifolium Ridl.
Dendrobium signatum Rchb.f.
Dendrobium spegidoglossum Rchb.f.
Dendrobium strongylanthum Rchb.f.
Dendrobium stuartii F.M.Bailey
Dendrobium stuposum Lindl.
Dendrobium sulcatum Lindl.
Dendrobium sutepense Rolfe ex Downie
Dendrobium terminale Parish & Rchb.f.
Dendrobium tetrodon Rchb.f. ex Lindl.
Dendrobium thyrsiflorum Rchb.f. ex Andre
Dendrobium tortile Lindl.
Dendrobium trigonopus Rchb.f.
Dendrobium truncatum Lindl.
Dendrobium umbonatum Seidenf.
Dendrobium unicum Seidenf.
Dendrobium venustum Teijsm. & Binn.
Dendrobium virgineum Rchb.f.
Dendrobium wardianum Warner

Dendrobium wattii (Hook.f.) Rchb.f.
Dendrobium williamsonii Day & Rchb.f.
Dendrobium wilmsianum Schltr.
Dendrobium xanthophlebium Lindl.
Dendrobium ypsilon Seidenf.

TONGA / TONGA (LES) / TONGA

Dendrobium calcaratum A.Rich.
Dendrobium calcaratum subsp. calcaratum
Dendrobium tokai Rchb.f.

TRINIDAD AND TOBAGO / TRINITÉ-ET-TOBAGO (LA) / TRINIDAD Y TOBAGO

Encyclia bradfordii (Griseb.) Carnevali & Ramírez
Encyclia fragrans (Sw.) Lemée

TURKS AND CAICOS

Encyclia rufa (Lindl.) Britton & Millsp.

UGANDA / OUGANDA (L') / UGANDA

Disa aconitoides Sond.
Disa aconitoides subsp. aconitoides
Disa aconitoides subsp. goetzeana (Kraenzl.) H.P.Linder
Disa eminii Kraenzl.
Disa erubescens Rendle
Disa erubescens subsp. erubescens
Disa fragrans Schltr.
Disa fragrans subsp. deckenii (Rchb.f.) H.P.Linder
Disa fragrans subsp. fragrans
Disa hircicornis Rchb.f.
Disa ochrostachya Rchb.f.
Disa scutellifera A.Rich.
Disa stairsii Kraenzl.
Disa welwitschii Rchb.f.
Disa welwitschii subsp. occultans (Schltr.) H.P.Linder
Disa welwitschii subsp. welwitschii

UNITED STATES OF AMERICA (THE) / ETATS-UNIS D'AMÉRIQUE (LES) / ESTADOS UNIDOS DE AMÉRICA (LOS)

Dendrobium macrophyllum A.Rich.
Dendrobium macrophyllum var. macrophyllum
Encyclia boothiana (Lindl.) Dressler
Encyclia boothiana subsp. boothiana
Encyclia cochleata (L.) Lemée
Encyclia pygmaea (Hook.) Dressler
Encyclia tampensis (Lindl.) Small

VANUATU / VANUATU / VANUATU

Dendrobium calcaratum A.Rich.
Dendrobium calcaratum subsp. calcaratum
Dendrobium capituliflorum Rolfe
Dendrobium conanthum Schltr.
Dendrobium delicatulum Kraenzl.
Dendrobium delicatulum subsp. delicatulum
Dendrobium erosum (Blume) Lindl.
Dendrobium goldfinchii F.Muell.
Dendrobium laevifolium Stapf
Dendrobium macranthum A.Rich.
Dendrobium macrophyllum A.Rich.
Dendrobium macrophyllum var. macrophyllum
Dendrobium masarangense Schltr
Dendrobium masarangense subsp. masarangense
Dendrobium mohlianum Rchb.f.
Dendrobium mooreanum Lindl.
Dendrobium polysema Schltr.
Dendrobium pseudorarum Dauncey
Dendrobium pseudorarum var. baciforme Dauncey
Dendrobium pseudorarum var. pseudorarum

VENEZUELA / VENEZUELA (LE) / VENEZUELA

Encyclia acuta Schltr.
Encyclia amicta (Lindl. & Rchb.f.) Schltr.
Encyclia angustiloba Schltr.
Encyclia arminii (Rchb.f.) Carnevali & Ramírez
Encyclia auyantepuiensis Carnevali & Ramírez
Encyclia bradfordii (Griseb.) Carnevali & Ramírez
Encyclia calamaria (Lindl.) Pabst
Encyclia ceratistes (Lindl.) Schltr.
Encyclia chacaoensis (Rchb.f.) Dressler & G.E.Pollard
Encyclia chimborazoensis (Schltr.) Dressler
Encyclia cochleata (L.) Lemée
Encyclia conchaechila (Barb.Rodr.)
Encyclia cordigera (Kunth) Dressler
Encyclia diurna (Jacq.) Schltr.
Encyclia fragrans (Sw.) Lemée
Encyclia garciniana (Garay & Dunst.) Carnevali & Ramírez
Encyclia grammatoglossa (Rchb.f.) Dressler
Encyclia gravida (Lindl.) Schltr.
Encyclia hartwegii (Lindl.) R.Vásquez & Dodson
Encyclia ionophlebium (Rchb.f.) Dressler
Encyclia ivonae Carnevali & G.A.Romero
Encyclia jauana Carnevali & Ramírez
Encyclia latipetala (C.Schweinf.) Pabst
Encyclia leucantha Schltr.
Encyclia lindenii (Lindl.) Carnevali & Ramírez

Encyclia livida (Lindl.) Dressler
Encyclia luteorosea (A.Rich. & H.Galeotti) Dressler & Pollard
Encyclia peraltense (Ames) Withner
Encyclia radiata (Lindl.) Dressler
Encyclia recurvata Schltr.
Encyclia sceptra (Lindl.) Carnevali & Ramírez
Encyclia sclerocladia (Lindl. ex Rchb.f.) Hoehne
Encyclia tigrina (Linden ex Lindl.) Carnevali & Ramírez
Encyclia venezuelana (Schltr.) Schltr.
Encyclia vespa (Vell.) Dressler

VIET NAM / VIET NAM (LE) / VIET NAM

Cymbidium aloifolium (L.) Sw.
Cymbidium atropurpureum (Lindl.) Rolfe
Cymbidium banaense Gagnep.
Cymbidium bicolor Lindl.
Cymbidium bicolor subsp. obtusum Du Puy & P.J.Cribb
Cymbidium cyperifolium Wall. ex Lindl.
Cymbidium cyperifolium subsp. cyperfolium
Cymbidium cyperifolium subsp. arrogans Du Puy & P.J.Cribb
Cymbidium dayanum Rchb.f.
Cymbidium devonianum Paxton
Cymbidium ensifolium (L.) Sw.
Cymbidium ensifolium subsp. ensifolium (L.) Sw.
Cymbidium erythrostylum Rolfe
Cymbidium finlaysonianum Lindl.
Cymbidium floribundum Lindl.
Cymbidium hookerianum Rchb.f.
Cymbidium insigne Rolfe
Cymbidium iridioides D.Don
Cymbidium lancifolium Hook.
Cymbidium lowianum Rchb.f.
Cymbidium lowianum subsp. lowianum Rchb.f.
Cymbidium macrorhizon Lindl.
Cymbidium sanderae (Rolfe) P.J.Cribb & Du Puy
Cymbidium schroederi Rolfe
Cymbidium sinense (G.Jacks. & H.C.Andr.) Wild.
Cymbidium suavissimum Sander ex Curtis
Dendrobium acinaciforme Roxb.
Dendrobium aduncum Wall. ex Lindl.
Dendrobium aloifolium (Blume) Rchb.f.
Dendrobium amabile (Lour.) O'Brien
Dendrobium annamense Rolfe
Dendrobium anosmum Lindl.
Dendrobium aphyllum (Roxb.) Fischer
Dendrobium bellatulum Rolfe
Dendrobium bilobulatum Seidenf.
Dendrobium brymerianum Rchb.f.
Dendrobium brymerianum var. brymerianum
Dendrobium capillipes Rchb.f.

Dendrobium cariniferum Rchb.f.
Dendrobium caryicola Guill.
Dendrobium chlorostylum Gagnep.
Dendrobium christyanum Rchb.f.
Dendrobium chrysanthum Wall.
Dendrobium chryseum Rolfe
Dendrobium chrysotoxum Lindl.
Dendrobium crepidatum Lindl. & Paxton
Dendrobium cretaceum Lindl.
Dendrobium crumenatum Sw.
Dendrobium crystallinum Rchb.f.
Dendrobium cumulatum Lindl.
Dendrobium dalatense Gagnep.
Dendrobium dantaniense Guill.
Dendrobium daoense Gagnep.
Dendrobium delacourii Guill.
Dendrobium densiflorum Wall. ex Lindl.
Dendrobium dentatum Seidenf.
Dendrobium devonianum Paxton
Dendrobium draconis Rchb.f.
Dendrobium exile Schltr.
Dendrobium farmeri Paxton
Dendrobium filicaule Gagnep.
Dendrobium fimbriatum Hook.
Dendrobium gratiosissimum Rchb.f.
Dendrobium hainanense Rolfe
Dendrobium hamatum Rolfe
Dendrobium harveyanum Rchb.f.
Dendrobium hemimelanoglossum Guill.
Dendrobium hendersonii A.D.Hawkes & A.H. Heller
Dendrobium hercoglossum Rchb.f.
Dendrobium heterocarpum Lindl.
Dendrobium incurvum Lindl.
Dendrobium indivisum (Blume) Miq.
Dendrobium indivisum var. indivisum (Blume) Miq.
Dendrobium intricatum Gagnep.
Dendrobium langbianense Gagnep.
Dendrobium leonis (Lindl.) Rchb.f.
Dendrobium lindleyi Steud.
Dendrobium linguella Rchb.f.
Dendrobium lituiflorum Lindl.
Dendrobium loddigesii Rolfe
Dendrobium lomatochilum Seidenf.
Dendrobium longicornu Wall. ex Lindl.
Dendrobium mannii Ridl.
Dendrobium moschatum (Buch.-Ham.) Sw.
Dendrobium nathanielis Rchb.f.
Dendrobium nobile Lindl.
Dendrobium nobile var. alboluteum Huyen & Aver.
Dendrobium nobile var. nobile
Dendrobium ochraceum De Wild.

Dendrobium ochreatum Lindl.
Dendrobium oligophyllum Gagnep.
Dendrobium Pachyglossum Par. & Rchb.f.
Dendrobium palpebrae Lindl.
Dendrobium parciflorum Rchb.f. ex Lindl.
Dendrobium parcum Rchb.f.
Dendrobium parishii Rchb.f.
Dendrobium pendulum Roxb.
Dendrobium perulatum Gagnep.
Dendrobium podagraria Hook.f.
Dendrobium porphyrophyllum Guill.
Dendrobium primulinum Lindl.
Dendrobium pseudointricatum Guill.
Dendrobium pseudotenellum Guill.
Dendrobium pulchellum Roxb. ex Lindl.
Dendrobium secundum (Blume) Lindl. ex Wall.
Dendrobium simondii Gagnep.
Dendrobium stuartii F.M.Bailey
Dendrobium terminale Parish & Rchb.f.
Dendrobium thyrsiflorum Rchb.f. ex Andre
Dendrobium tortile Lindl.
Dendrobium truncatum Lindl.
Dendrobium unicum Seidenf.
Dendrobium venustum Teijsm. & Binn.
Dendrobium virgineum Rchb.f.
Dendrobium wardianum Warner
Dendrobium wattii (Hook.f.) Rchb.f.
Dendrobium williansonii Day & Rchb.f.

YEMEN / YÉMEN (LE) / YEMEN (EL)

Disa pulchella Hochst. ex A.Rich.

ZAMBIA / ZAMBIE (LA) / ZAMBIA

Disa aconitoides Sond.
Disa aconitoides subsp. aconitoides
Disa aconitoides subsp. concinna (N.E.Br.) H.P.Linder
Disa aequiloba Summerh.
Disa aperta N.E.Br.
Disa caffra Bolus
Disa celata Summerh.
Disa cryptantha Summerh.
Disa dichroa Summerh.
Disa eminii Kraenzl.
Disa englerana Kraenzl.
Disa equestris Rchb.f.
Disa erubescens Rendle
Disa erubescens subsp. carsonii (N.E.Br.) H.P.Linder
Disa erubescens subsp. erubescens
Disa hircicornis Rchb.f.

Disa katangensis De Wild.
Disa miniata Summerh.
Disa nyikensis H.P.Linder
Disa ochrostachya Rchb.f.
Disa ornithantha Schltr.
Disa perplexa H.P.Linder
Disa robusta N.E.Br.
Disa roeperocharoides Kraenzl.
Disa satyriopsis Kraenzl.
Disa saxicola Schltr.
Disa similis Summerh.
Disa ukingensis Schltr.
Disa verdickii De Wild.
Disa walleri Rchb.f.
Disa welwitschii Rchb.f.
Disa welwitschii subsp. occultans (Schltr.) H.P.Linder
Disa welwitschii subsp. welwitschii
Disa zombica N.E.Br.

ZIMBABWE / ZIMBABWE (LE) / ZIMBABWE

Disa aconitoides Sond.
Disa aconitoides subsp. aconitoides
Disa aconitoides subsp. concinna (N.E.Br.) H.P.Linder
Disa cornuta (L.) Sw.
Disa equestris Rchb.f.
Disa erubescens Rendle
Disa erubescens subsp. erubescens
Disa fragrans Schltr.
Disa fragrans subsp. fragrans
Disa hircicornis Rchb.f.
Disa miniata Summerh.
Disa ochrostachya Rchb.f.
Disa ornithantha Schltr.
Disa patula Sond.
Disa patula var. transvaalensis Summerh.
Disa patula var. patula
Disa perplexa H.P.Linder
Disa polygonoides Lindl.
Disa rhodantha Schltr.
Disa rungweensis Schltr.
Disa saxicola Schltr.
Disa versicolor Rchb.f.
Disa walleri Rchb.f.
Disa welwitschii Rchb.f.
Disa welwitschii subsp. welwitschii
Disa woodii Schltr.
Disa zimbabweensis H.P.Linder
Disa zombica N.E.Br.